Tonopah

The Greatest, the Richest, and the Best Mining Camp in the World

Robert D. McCracken

Nye County Press

TONOPAH NEVADA

TONOPAH
The Greatest, the Richest, and the Best Mining Camp in the World*
by Robert D. McCracken

* This description of Tonopah appeared in the *Tonopah Bonanza,* April 29, 1905, p. 8.

Second printing 1992
© Copyright 1990 by Nye County Press

Published in 1990 by Nye County Press
P.O. Box 3070
Tonopah, Nevada 89049

Library of Congress Catalog Card Number: 90-060549
ISBN: 1-878138-50-2

DESIGNED BY PAUL CIRAC, WHITE SAGE STUDIOS, VIRGINIA CITY, NEVADA
PRINTED IN THE UNITED STATES OF AMERICA

Tonopah

The Greatest, the Richest, and the Best Mining Camp in the World

To my father, Robert G. McCracken,
and all the other miners who
worked underground in Nye County

To the Metscher brothers
for their tireless efforts
to understand and preserve
Nevada history

In appreciation for their unwavering support and encouragement for the Nye County Town History Project:

Nye County Commissioners

Robert "Bobby" N. Revert
Joe S. Garcia, Jr.
Richard L. Carver
Barbara J. Raper

and Nye County Planning Consultant

Stephen T. Bradhurst, Jr.

Contents

Preface

Historians generally consider the year 1890 as the close of the American frontier. By then, most of the western United States had been settled, ranches and farms developed, communities established, and roads and railroads constructed. The mining boomtowns, based on the lure of overnight riches from newly developed lodes, were but a memory.

Although Nevada was granted statehood in 1864, examination of any map of the state from the late 1800s shows that although much of the state was mapped and its geographical features named, a vast region — stretching from Belmont south to the Las Vegas meadows, comprising most of Nye County— remained largely unsettled and unmapped. In 1890 most of southcentral Nevada remained very much a frontier, and it continued to be so for at least another twenty years.

The great mining booms at Tonopah (1900), Goldfield (1902), and Rhyolite (1904) represent the last major flowering of what might be called the Old West. Consequently, southcentral Nevada, notably Nye County—perhaps more than any other region of the West—remains close to the American frontier. In a real sense, a significant part of the frontier can still be found there. It exists in the attitudes, values, lifestyles, and memories of area residents. The frontier-like character of the area also is visible in the relatively undisturbed condition of the natural environment, most of it essentially untouched by humans.

Aware of Nye County's close ties to our nation's frontier past and the scarcity of written sources on local history (especially after 1920), the Nye County Board of Commissioners initiated the Nye County Town History Project (NCTHP) in 1987. The NCTHP is an effort to systematically collect and preserve the history of Nye County. The centerpiece of the NCTHP is a large set of interviews conducted with individuals who had knowledge of local history. The interviews provide a composite view of community and county history, revealing the flow of life and events for a part of Nevada that has heretofore been largely neglected by historians. Each interview was recorded, transcribed, and then edited lightly to preserve the language and speech patterns of those interviews. All oral history interviews have been printed on acid-free paper and bound and archived in Nye County libraries, Special Collections in the James R.

Dickinson Library at the University of Nevada, Las Vegas, and at other archival sites located throughout Nevada.

Collection of the oral histories has been accompanied by the assembling of a set of photographs depicting each community's history. These pictures have been obtained from participants in the oral history interviews and other present and past Nye County residents. Complete sets of these photographs have been archived along with the oral histories.

The oral histories and photo collections, as well as written sources, served as the basis for the preparation of this volume on Tonopah history. It is one in a series of volumes on the history of all major Nye County communities.

In a real sense this volume, like the others in the NCTHP series, is the result of a community effort. Before the oral interviews were conducted, a number of local residents provided advice on which community members had lived in the area the longest, possessed and recalled information not available to others, and were available and willing to participate. Because of time and budgetary constraints, many highly qualified persons were not interviewed.

Following the interviews, the participants gave even more of their time and energy: They elaborated upon and clarified points made during the taped interviews; they went through family albums and identified photographs; and they located books, dates, family records, and so forth. During the preparation of this manuscript, a number of community members were contacted, sometimes repeatedly (if asked, some would probably readily admit that they felt pestered), to answer questions that arose during the writing and editing of the manuscript. Moreover, once the manuscripts were in more or less final form, each individual who was discussed for more than a paragraph or two in the text was provided with a copy of his or her portion of the text and was asked to check that portion for errors. Appropriate changes were then made in the manuscript.

Once that stage was completed, several individuals in Tonopah were asked to review the entire manuscript for errors of omission and commission. At each stage, this quality-control process resulted in the elimination of factual errors and raised our confidence in the validity of the contents.

The author's training as an anthropologist, not a historian (although the difference between the disciplines is probably less than some might suppose), likely has something to do with the community approach taken in the preparation of this volume. It also may contribute to the focus on the details of individuals and their families as opposed to a general description of local residents and their communities. Perhaps this volume, as well as a concern with variability among individuals and their contribution to a community, reflects an "ethnographic," as opposed to a "historical," perspective on local history. In the author's view, there is no such thing as "the history" of a community; there are many histories of a community. A community's history is like a sunrise — the colors are determined by a multitude of factors, such as the time of year, weather, and point of view. This history of Tonopah was greatly determined by the input of those who helped produce it. If others had participated, both the subjects treated and the relative emphasis the subjects received would have been, at least, somewhat different. Many basic facts would, of course, remain much the same—such things as names, dates, and locations of events. But the focus, the details illustrating how facts and human beings come together, would have been different. History is, and always will remain, sensitive to perspective and impressionistic, in the finest and most beautiful sense of the word.

A longer and more thoroughly referenced (though non-illustrated) companion to this volume, titled *A History of Tonopah, Nevada,* is also available through Nye County Press. Virtually all written material contained in the present volume was obtained from the longer volume. Those who desire more comprehensive referencing should consult the longer version of Tonopah history.

I hope that readers enjoy this illustrated history of Tonopah, Nevada. Tonopah is a very special and interesting place — part frontier, part modern; part yesterday, part tomorrow. Situated in the high desert on some of the most beautiful real estate in the world, it was one of the most notable mining camps in history. If not "The Greatest, the Richest and the Best Mining Camp in the World," it was certainly in the running for that title.

Robert D. McCracken

Acknowledgments

This volume was produced under the Nye County Town History Project, initiated by the Nye County Board of Commissioners. Appreciation goes to Chairman Joe S. Garcia, Jr., Robert "Bobby" N. Revert, and Pat Mankins; Mr. Revert and Mr. Garcia, in particular, showed deep interest and unyielding support for the project from its inception. Thanks also go to current commissioners Richard L. Carver and Barbara J. Raper, who have since joined Mr. Revert on the board and who have continued the project with enthusiastic support. Stephen T. Bradhurst, Jr., planning consultant for Nye County, gave unwavering support and advocacy, provided advice and input regarding the conduct of the research, and constantly served as a sounding board as production problems were worked out. This volume would never have been possible without the enthusiastic support of the Nye County commissioners and Mr. Bradhurst.

Thanks go to the participants of the Nye County Town History Project, especially those from Tonopah, who kindly provided much of the information; thanks, also, to residents from Tonopah and throughout southern Nevada — too numerous to mention by name — who provided assistance, historical information, and photographs, many of which are included in this volume.

Jean Charney and Jean Stoess did the word processing and, along with Gary Roberts, Maire Hayes, and Jodie Hanson, provided editorial comments, review, and suggestions. Alice Levine and Michelle Starika edited several drafts of the manuscript and contributed measurably to this volume's scholarship and readability; Alice Levine also served as production consultant. Gretchen Loeffler and Bambi McCracken assisted in numerous secretarial and clerical duties. Gordon Loeffler and Donn Knepp copied photographs; Paul Cirac, who was raised in central Nevada, was responsible for the design and layout of this book.

William J. Metscher, who probably knows more about Tonopah history than any other individual, helped select and identify all photographs presented here from the Central Nevada Historical Society's collection on Tonopah. He also kindly critiqued several drafts of the manuscript, and his assistance and support have been invaluable. Albert N. Bradshaw, Norman Coombs, Don B. Potts, Jeanne Potts, Edward R. Slavin, and Solan Terrell also made thoughtful comments and graciously answered many questions regarding local history. Kevin Rafferty and Lynda Blair, from the University of Nevada, Las Vegas, Environmental Research Center, provided helpful suggestions on the section concerning the archaeology of Native Americans in the Tonopah area; Margaret J. Waski, with the Tonopah Resource Area, U.S.D.I., Bureau of Land Management, also provided advice on archaeology. Phillip Earl of the Nevada Historical Society contributed valuable support and criticism throughout, and Tom King at the Oral History Program of the University of Nevada, Reno, served as consulting oral historian. Susan Jarvis of Special Collections, James R. Dickinson Library, University of Nevada, Las Vegas, assisted greatly with research conducted at that institution. Much deserved thanks are extended to all these persons.

All aspects of production of this volume were supported by the U.S. Department of Energy, Grant No. DE-FG08-89NV10820. However, any opinions, findings, conclusions, or recommendations expressed herein are those of the author and do not necessarily reflect the views of DOE. Any errors and deficiencies are, of course, the author's responsibility.

R. D. M.

Tonopah

The Greatest, the Richest, and the
Best Mining Camp in the World

Famous photo (taken by Mimosa Pittman, wife of Senator Key Pittman) of a lightning strike in Tonopah, Nevada. (Copyright 1904 by Key Pittman.)

Introduction

Tonopah and the mining communities that it spawned in the southern part of Nevada — including Las Vegas, which arguably might be a far different place today were it not for Tonopah — represent the last flowering of the Old West in America. In the Old West, the discovery of precious metals and ranching combined with individualism and the lure of wealth and adventure to create communities immortalized in song and lore because they symbolize the core American values of personal freedom and opportunities for self-betterment. Perhaps more than almost any place in the country, much of the Old West still survives in the Tonopah region; this helps make central Nevada a special place — an unusually fascinating area.

The era of the underground miner has almost ended. This type of mining involves skills that few know and for which there now is almost no demand. As the world turns more and more to open-pit mines and to plastic and ceramic replacements for metals, there is no reason to think that demand will be revived. The era of the small-time mine operator and leaser has also vanished. To be a leaser one had to be a dreamer, a wishful thinker — one whose expectations, perhaps, had become confused with his hopes — a chaser of rainbows. In Tonopah the leasers started the camp and they closed it. Many made good money, but most made only wages, if that.

The glory days of Tonopah were sandwiched between the two eras of leasing. The town boomed. But companies dominated by wealthy Eastern capitalists controlled the mining and thus the economy. The Easterners were in mining for the money and cared little for the town or the miners and their families. They took what they could and then they left. They took the wealth from the hills and left the town to survive on its own.

But Tonopah did survive! Unlike so many mining camps in central Nevada, it made the transition from mining to a mixed economy. World War II brought the air base and, later, there was the influx of other defense-related money.

During the 1980s, the boom in open-pit mining, including the molybdenum/copper deposit north of town and extensive activities at Round Mountain and Candelaria, provided further stimulation for the Queen of the Silver Camps, once called "the greatest, the richest and the best mining camp in the world" (*Tonopah Bonanza*, April 29, 1905).

1

An Indian woman and her daughter in Tonopah, circa 1904. Note the water jug, made waterproof by a covering of pitch; a lard can serves as a flower pot.

Central Nevada Historical Society – UNLV Special Collections

CHAPTER ONE

The First Inhabitants

T he archaeological history of the Tonopah area is divided into two major epochs — the Pre-Archaic and the Archaic. The Pre-Archaic period in the Tonopah area began about 10,000 years ago and lasted until about 6000 years ago, at which time the Archaic period began. The Archaic period has three subdivisions: Early Archaic (6000-1500 years ago); Middle Archaic (1500 years ago-A.D. 500); and Late Archaic (A.D. 500-about A.D. 1800).

There is little solid evidence of human beings in the New World before about 13,000 years ago. Prior to that time, much of the North American continent was covered by two massive collateral sheets of ice, each up to 7500 feet thick. The intensity of this most recent ice age waxed and waned, reached peaks at about 65,000 and 18,000 years ago, and ended dramatically about 12,000 to 9000 years ago. At their maxima, the ice sheets had incorporated so much of the planet's water that the level of the oceans fell by 300 or more feet, creating a land bridge between Asia and Alaska known as Beringia. The first human beings to occupy North America are assumed to have crossed over from Asia on the Beringia land bridge and to have moved southward to the northern plains.

There are few buried archaeological sites in the western Great Basin that represent Pre-Archaic people, and none in the central part of Nevada. Yet, stone implements belonging to the Pre-Archaic people, including large knives, projectile points with ground edges, crescent-shaped objects, gravers, punches, choppers, and several types of scrapers, have been found, usually near the marshy deltas of streams feeding shallow lakes, including Mud Lake south of Tonopah. These people probably hunted big game, including extinct species such as mammoths and mastodons, utilized smaller animals, and processed foods such as cattail shoots, pollen, and green seeds. It is thought that they were nomadic, that they did not construct permanent structures or store food, and that they did not grind seeds. This lifestyle was made possible by the cooler and more moist conditions that prevailed at the time.

During the Archaic period there were great changes in the lives of the early inhabitants of central Nevada. Settlement patterns changed; people adopted a home range or a sequence of home ranges varying with the season — winter camps, seasonal base camps, work sites, and so forth. People were probably sedentary in the winter, building substantial shelters and storage facilities. Archaic peoples participated in game drives and ambushes, using rock walls, lines of cairns, or brush fences. During the Archaic period we find the first rock art, which may have been related to hunting cults or perhaps served to mark territory.

The Early Archaic period was accompanied by a warming and drying trend in climate. It resulted in the shrinkage or disappearance of most lakes and marshes. Strange as it may seem, researchers believe that the piñon-juniper woodlands, so much a part of the present environment, probably did not appear in the western Great Basin until about 4000 B.C. The arrival of the piñon pine in central Nevada provided a new food source for the inhabitants. Remains of the Early Archaic peoples of the Tonopah area can be found in the Toquima Mountains and in Reese River Valley and Monitor Valley.

The climate during the Middle Archaic period was cool and moist, and food was consequently more readily available to the residents of central Nevada through the formation of meadows, marshes, and shallow lakes. Archaeological sites for the period are distinguished by the remains of large animals, including mountain sheep, antelope, deer, bison, and elk. Although big game hunting was important, remains of smaller animals, particularly rabbits, are found in greater abundance than previously.

The Late Archaic period was accompanied by a mild warming and drying trend. The atlatl, a spear-thrower, was replaced by the bow and arrow, and projectile points became smaller and lighter. Pottery was introduced after about A.D. 1100. Plant-processing equipment became more elaborate and abundant, and a wider variety of foods from a number of ecozones were utilized. Plant foods and small game, especially rabbits, were the order of the day, with a reduction in the importance of large game. Many of these changes may have also been produced by an increase in population. House size, settlement size, and sedentism all increased during this period, and greater reliance on the use of pine nuts is evident, anticipating the culture of the Western Shoshone Indians who occupied the central Nevada area at the time of European contact.

Many researchers believe that the Western Shoshone first migrated into the area approximately 1000 years ago from somewhere near the Death Valley area. Other authorities believe that the ancestors of the Western Shoshone speakers were living in the Great Basin 10,000 years ago and attained their historic distributions about 1000 years ago.

The Western Shoshone Indians

When Europeans first arrived in the Tonopah area (certainly by 1827, and perhaps earlier) they encountered people who had been living there for at least 1000 years. The Western Shoshone employed a markedly diverse strategy for existing in the central Great Basin, one which utilized a wide variety of foods. They were what anthropologists call hunters and gatherers. The basic social unit was the family, and family clusters foraged for food in small groups from the spring through the fall. Plants were the basis of the Shoshone economic system. Nuts and other seeds were the mainstay of the diet. It has been estimated that a typical Western Shoshone family of four could gather approximately 1200 pounds of pine nuts each autumn. This amount would have lasted a family about four months. During the spring and summer, the families tended to move to where plants and seeds were abundant. Baskets coated with pine pitch were used to transport water.

In addition to plant foods, the Western Shoshone relied upon hunting of large and small game. Bighorn sheep were important, and in the summer their movements were monitored and then, with the aid of dogs, they were ambushed at a predesignated spot. During the fall, hunters intercepted the bighorn from behind rock walls and hunting blinds. Rabbits were also important and were hunted in communal drives involving men and women. Burrowing rodents, including pocket gophers and ground squirrels, were either flooded or smoked out or

Indian women shelling pine nuts at the cattle corrals above Tonopah. Picture taken prior to 1940.

dug out with rodent skewers. The Western Shoshone also hunted doves, mockingbirds, sage hen, quail and — less frequently — owl, hawk, and crow. When available, large black Mormon crickets in swarms were utilized. Labor was divided according to sex, with the women gathering plant foods and the men hunting.

During the winters the Western Shoshone occupied traditional winter villages, especially in the Reese River, Ruby, Spring and Big Smoky valleys, where resources were plentiful. Such "villages," however, did not have permanent structures. The typical winter hut, which provided shelter for a family of six, would consist of a light frame covered with slabs of bark and surrounded with a tier of stones to keep the supports firmly planted.

Clothing varied according to age, sex, and the individual hunter's ability. Children often went naked. Even in subzero temperatures clothing was often scarce. A sewed robe, most often

The Greatest, the Richest, and the Best Mining Camp in the World

Tim Hooper on the left, Jennie Crow Hooper holding baby Albert Hooper. August 15, 1902, Fallon, Nevada, area. Albert Hooper and his wife, Mayme, resided for more than 50 years in the Tonopah area.

Nye County Town History Project – Hooper Collection

made of rabbit skin, was the most common protection against the cold. Sheep, antelope, and deer hides were also worn. Women preferred long gowns of skin in the winter, and when skins were scarce they wore skirts of bark or grass.

In their religion, the Western Shoshone engaged in a direct relationship with the supernatural; there was no priesthood, but supernatural powers could be obtained through visions and dreams. Ailments were treated with plants, including 52 different plant remedies for "colds," 57 for venereal diseases, 44 for "swellings," 34 for diarrhea, 37 for rheumatism, and 48 for various stomach ailments. Bodies of the dead were treated variably; sometimes cremated, sometimes abandoned or burned in their dwellings. In mountain areas they were often buried in rockslides or talus slopes.

Within the Tonopah area, Western Shoshone camps existed at Stonewall Mountain; Breen Creek, Horse Canyon, and Rose's Spring on the west side of the Kawich Range; the Reveille Mill on the east side of the Kawich; Stone Cabin; Tybo; Warm Springs; Hot Creek; Twin Springs; Moray; Duckwater; Darrough's Hot Springs in Big Smoky Valley; Peavine Creek; and at other localities, especially at numerous sites on the west side of the Toiyabe Range and the east side of the Shoshone Mountains along the Reese River Valley.

Indians After European Contact

The arrival of Europeans in the Western Shoshone territories following the discovery of the famed Comstock Lode in western Nevada, as well as mineral discoveries in the Reese River Valley and the Toiyabe and Toquima mountains a few years later, severely disrupted the traditional way of life for the Western Shoshone. Mining led to loss of land, conflicts over water, disappearance of game, and the logging of piñon trees for charcoal production, mine timbers, posts, and fuel. Traditional seed-collecting areas were sometimes plowed under. Mining led to the deterioration of streams and the spread of tailings over useful areas. These changes produced disruptions in the traditional economic activities of the native Americans. The Indian people responded in large part by forming settlements on the outskirts of mining camps, railroad towns and farming communities or by attaching themselves to local ranches. Traditional activities were abandoned; men hauled and chopped firewood, sold pine nuts and fish, hauled water, dug irrigation ditches, worked as loggers, plowed fields, hunted rabbits for bounty, or hunted large game commercially. Women worked as laundresses, maids, and kitchen helpers.

Conflicts between Indians and whites sometimes occurred, and the government responded by establishing treaties and setting aside reserves of land for the Indians. However, the government did not always keep its promises regarding the reserves. Native American populations reached a low in the last years of the nineteenth century due primarily to malnourishment and the ravages of European diseases introduced over the previous decades.

Early Exploration

In 1826, Jedediah Smith became the first white man on record to travel overland to California. He made his way down the Colorado River along Nevada's southern border. In May 1827, on the return leg of his journey, after being the first white man to cross the Sierra Nevada, he entered the present state of Nevada just south of Walker Lake. From there he

> moved between the Gabbs Valley and Pilot Ranges, around the southern end of the Shoshone and Toiyabe Ranges and crossed Big Smoky Valley just south of Peavine Creek. He followed a well-marked Indian trail across Big Smoky Valley ... then crossed the southern tip of the Toquima Range directly east from Manhattan ... camped for two nights, June 7 and 8, near the site of Belmont (Thomas, 1982:7-8).

The Canadian trapper-explorer Peter Skene Ogden was probably the second white known to enter what is now Nye County and the Tonopah area. As a young man, Ogden was employed as a clerk for the Northwest Fur Trading Company. He eventually became leader of the Hudson's Bay Company's Snake Country Expedition. As

Peter Skene Ogden was probably the second white man to enter what is now Nye County.

University of Nevada, Las Vegas – Dickinson Library Special Collections

leader, he became the first white man to travel down and record the Humboldt River. In his 1829-1830 expedition, Ogden followed the river downstream to the Humboldt Sink, then south to the Carson Sink, and on to the Walker River. South of Walker Lake the expedition traveled "parallel to the present Nevada-California border, eventually reaching the Colorado River drainage basin and the Gulf of California before heading back to Oregon" (Funk, 1982:7L). Peter Ogden later wrote of this trip across the Nevada desert:

> There were times when we tasted no food, and were unable to discover water for several days together; without wood, we keenly felt the cold; wanting grass, our horses were reduced to great weakness, so that many of them died, on whose emaciated carcasses we were constrained to satisfy the intolerable craving of our hunger, and as a last resort, to quench our thirst with their blood (Cline, 1974:91-93).

John C. Fremont crossed southern Nevada on three occasions, each time entering Nye County. In November 1845, Fremont traveled through the Great Smoky Valley. He is said to have given the valley the name Big Smoky because of the blue haze that resembles smoke hanging over the valley and surrounding mountains, especially in the fall. On his way through the valley, Fremont marked San Antonio Peak, Hot Springs, Twin Rivers, and Smoky Creek. On November 14 he camped at a small creek that he called Basin Creek, later known as Big Smoky Creek and Kingston. On November 16, Fremont camped at a boiling hot springs now known as Darrough's Hot Springs.

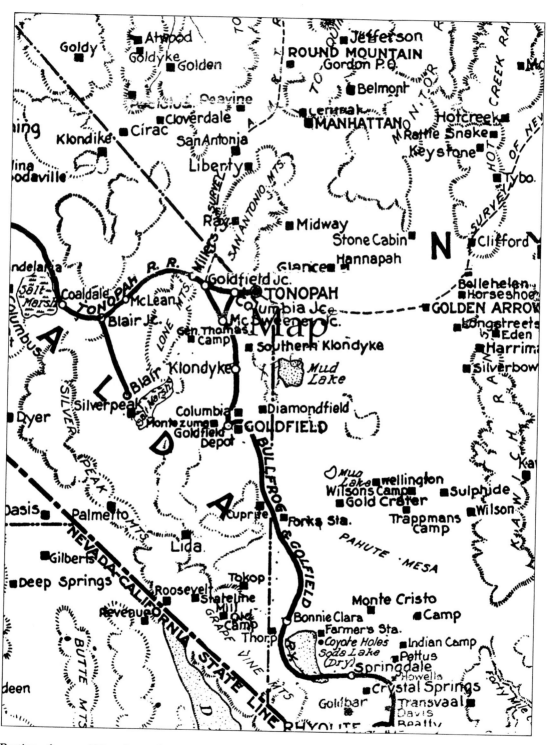

Portion of map of Nevada produced by the Clason Map Company, Denver, Colorado, in 1905. Note the many mining camps in the Tonopah area.

CHAPTER TWO

The Search for Silver

E very mining camp was replete with tales of the near-miss of the lucky strike — the "one that got away." "I was inches away from a rich strike," or "I walked right over ground that later produced millions" were common laments. The rich silver ore in Tonopah literally stuck out of the ground, and any number of individuals were close and could have made the discovery that Jim Butler made in 1900.

Silver Peak, located about 30 miles southwest of Tonopah, was organized as a mining district in 1865, and there was an active post office there from 1866 to 1913. The town of Candelaria, 25 miles west of Tonopah, dates to 1875 and was a booming mining district in 1880. Prospectors at Silver Peak fanned out in all directions in search of fruitful signs of other mineral deposits. They were, of course, familiar with the sawtooth peak in the San Antonio Range, which separated Ralston Valley on the east and San Antonio Valley to the west. Although there were many passes across the San Antonio Range, the lowest and least rough took them along the base of Sawtooth Peak, now known as Mount Butler. There was an especially good trail connecting the station at Stone Cabin about 40 miles east of Tonopah and the cattle ranges along Lower Peavine Creek in the San Antonio Valley, extending westerly toward Silver Peak and Candelaria.

Many cowboys and travelers must have seen the 300-yard-square outcropping on the trail through Sawtooth Pass (later known as Tonopah Pass), but it was black or brownish black and in southern Nevada such outcroppings typically ran high in iron but low in everything else except silica. The cowboys and prospectors must have ignored the black float and the black ledges, thinking the color was due to iron compounds rather than compounds of manganese and silver, the true coloring agents of the rock. Moreover, the springs located three miles from the outcropping, later called Tonopah Springs, were known as a rendezvous for herders; many of them saw the ledges that ran east and west across the face of Mount Oddie down to the flat.

Sometime in the 1890s an old man from Silver Peak made several trips toward Sawtooth Peak and reported he had located some ledges of black quartz near its base. Many supposed that he must have made a valuable discovery, for he allowed no one to accompany him on these trips. He failed to return from one of his trips and was never seen again. It is not known whether

he lost his life on the desert or whether he simply moved on to another, more attractive area. He was a close-mouthed individual, and he never discussed his plans. What little he did say, however, fits very closely with the ledges at Tonopah.

An Indian known as Charlie Fishman was the next person to almost find Tonopah's fabulous treasure. Fishman is said to have told the assayer at the Silver Peak Mine that he knew the location of some "big, black quartz ledges." He said they looked good and that they might contain gold. Fishman was intelligent, knew the art of prospecting, and had made previous prospecting trips on horseback. The assayer, known as "Van" to the whites and "Mr. Van" to the Indians, was interested in prospecting and encouraged it to the extent that his finances allowed. Van asked Fishman how long it would take to make a trip to examine the ledges for gold. Fishman replied that it would take two to three weeks if he had an outfit. The assayer supplied Fishman with a light wagon, a team of horses, and enough supplies for three weeks; the Indian said that he knew where water could be obtained. Van instructed Fishman to pan for gold along all the ledges. Fishman returned in approximately three weeks and reported that he had panned the ledges and found but one color.

Unfortunately for both men, Fishman failed to bring back any of the rock. He returned what was left of the grubstake and disappeared. In 1901 Van paid a visit to the new discoveries at Tonopah. He had heard tales of the mineral wealth, the scores of teamsters and hundreds of horses and the many leasers in the camp. He is said to have inspected the leases on Mizpah Hill and then to have crossed over to the Valley View Hill. As he stood on the edge of the first lease

(Facing page) One of the earliest photos of Tonopah, taken in 1901, shortly after Jim Butler began letting leasers work on the claims he and his wife, Belle, had staked. Belle Butler is with the dog on the left; behind her are sacks of ore, which the miners had taken from veins just below the grassroots, at the approximate site of the Mizpah Mine. Mount Oddie is in the background.
Central Nevada Historical Society – Ira Jacobsen Collection

(right) A few Blacks resided in Tonopah during its early days. This unidentified man and woman are standing before what appears to be an adobe dugout. Bottles and tin cans have been stacked to form cribbing.
Central Nevada Historical Society – Ira Jacobsen Collection

he came to on Valley View Hill, he looked down into an open cut where ore was being broken and spied Fishman drilling a hole with steel and a single jack. He called to the Indian and Fishman looked up, saying, "Hello, Mr. Van, how are you? This is the place where I found the black quartz." Without a word Van turned away, walked down the hill, hitched up his team and left Tonopah, never to return. So close, but so far away (Keeler, 1913:965).

Warren Averill's Springs, located a short distance north of Tonopah, served as a rendez-vous for herders for many years; those who camped there were probably familiar with the ledges at Sawtooth Pass. Warren Averill, an old freighter for whom the springs were named, wintered his oxen at the site for a period of years prior to 1900. Jim Butler later changed the springs' name from Warren Averill's Springs to Tonopah Springs, noting that the Indians used to designate the place by a Numic word that means "brush water springs." In Shoshone, the word would be properly spelled as *lonobe-bah.* (More recently it has been suggested that the word should be *tonobe-bah,* as there is not an "l" in the Shoshone language at the beginning of words.) Carlson notes:

> The local Indian word *tonopah* has been interpreted as "hidden spring," "brush water spring," "greasewood spring," "little water" and "water brush," the last being accepted by old-timers in the area ... The name appears to mean "greasewood water (spring)" and to derive from either Shoshone (Central Numic) *to-nuv,* "greasewood," or Northern Paiute (Western Numic) *to-nav,* "greasewood," and *pa,* "water," in both dialects ... (1974:233-234).

The Greatest, the Richest, and the Best Mining Camp in the World

In 1901 leases were being worked and were more developed with small headframes built over the shafts. Ore was hoisted out of the shafts by means of a whim that was powered by the horse at the left. From left are Cal Brougher, Jim Butler, Belle Butler, E.W. Smith, who took the photograph, a group of leasers, and Tasker Oddie at the far right.

The Big Strike!

T he story of Jim Butler's discovery is well known and is only briefly reviewed here. Butler had come from California and established a ranch in Monitor Valley. He was married to Belle, a woman he met while in Tybo. Fluent in Shoshone, he loved to talk to Indians. In May 1900, leaving Belle at home and traveling alone, Butler ostensibly went to visit the Bell and Court discovery in the Klondike district, about 14 miles south of Tonopah, which had been established the previous year. Instead of heading straight down Ralston Valley to Mud Lake (then called Cactus Lake) to the west edge of the Klondike Hills and then across to the Bell and Court property, Butler crossed the San Antonio Mountains at a high, rough pass to Tonopah Springs and then went to Sawtooth Peak.

Historians have since surmised that, although Butler may have been interested in visiting Klondike, he was also interested in prospecting and had a particular site in mind. Some believe that the Indians had told him of the ledges in Sawtooth Pass and that it was his intention from the beginning of the trip to check them out. Thus, the tale of Jim Butler camping in Sawtooth Pass and picking up a black stone to throw at his burro, then noting the unusual weight of the stone, is myth. At any rate, Butler collected the samples on May 19, 1900.

Butler took samples from the ledge at Sawtooth Pass, later known as the Mizpah Vein, and continued on to South Klondike. There he offered assayer Frank Hicks an interest in the discovery if he would assay the samples. Hicks is said to have replied that he would not give a dollar for a thousand tons of such stuff and to have thrown the samples on the dump. Butler is reported to have shown the ore to several other people who also did not think much of it. On May 25, 1900, Butler returned to Belmont by the same route, gathering more samples. Back in Belmont he showed them to the regulars at Wilse Brougher's store. Tasker L. Oddie was one of these, and he offered to have an assay made. Butler had no money, but his wife, Belle, offered Oddie an interest in the mine if the samples proved good. Oddie sent the samples to Walter Gayheart in Austin, who ran them and found them to be high in silver and gold. His assays ran from $18 to $600 per ton. Gayheart informed Oddie of the results and Oddie sent word to Butler.

Meanwhile Butler had returned to his ranch in Monitor Valley, where he ignored Oddie's urgent message for three weeks. Oddie sent copies of the assays by Indian runner to the ranch, but Butler was busy haying. When Butler did go to Belmont it was to concentrate on his official duties there as district attorney. In August 1900, Brougher and Oddie outfitted Jim and Belle Butler in Belmont and the couple returned to Sawtooth Pass to locate claims. News of the discovery had already leaked out, and there was a small rush to the Tonopah region before Butler and his wife returned in August. Because no one knew exactly where the samples had been obtained and Butler had not put up location notices, the valuable discoveries still had not been located. The first claim located by Butler was the Desert Queen, followed by the Burro,

Tonopah's First Cook

U ntil April 1901, Sadie Grieves, Belle Butler, and Charlotte Stimler were the only women in the Tonopah camp. Charlotte Stimler, known as Lottie Stimler Nay, has left an account of her first few months in Tonopah that graphically illustrates the hardships the first residents of Tonopah endured.

Lottie Stimler was born and raised in Belmont. During the last years of the nineteenth century, Belmont mining had suffered a slow decline, and with the discovery at Tonopah, most Belmont residents decamped for the new community. Lottie's brother Harry left to cook for the miners employed by Butler, Brougher, and Oddie. Lottie became discontented; she was tired of the dreary, monotonous life in Belmont. She longed for the adventure of a new, exciting town, so she sent for a 14x15-foot tent, some cooking utensils and dishes and, with Harry's help, she left Belmont on January 25, 1901, for Tonopah camp. At the camp, meals had been prepared over open fires and served to the miners who squatted on the ground or sat on bedrolls; so there was

Lottie Stimler Nay, April 1901.

great rejoicing and a general holiday declared when Miss Stimler drove into camp on January 29, 1902 [sic; sources differ on the year, but it had to be 1901], with her two wagons loaded with groceries and furniture. All the men assisted in the work of setting up the tents. About noon the wind began to rage, but everyone worked on, hoping to make it possible to use the tent that night (Sawyer, 1953:7-8).

Of her new quarters, Stimler said:

I thought I had never seen such a horrid place as that tent: everything was so crowded, dusty and dirty, and the water wagon had failed to get into camp before dinnertime. My "white" dishes were put to immediate use, as was also my table and my new white oilcloth covers. When all the men came in from work, great was their surprise and delight to find a decent table at which they could all sit down together without discomfort, and the

white dishes — how they enjoyed eating off of them, for they had used tin ones before ... That night at supper there were 30 men instead of about 15, as Miss Grieves and I had calculated; but fortunately I had cooked up a kettleful each of beans and sauerkraut that forenoon, and then we made biscuits and fried steak, and managed to "fill them up." The boys had had to fix the tents in such a hurry that the cold wind seemed to blow right through the dining room without stopping; and the men had a great time trying to keep the two lamps burning during the meal, but they were all good natured and seemed well pleased with the first supper (Doughty, 1974:3).

Stimler goes on to say, "About the second of February it began to snow and blow so furiously that within a few days the snow was three feet on the level and piled up behind our kitchen for five or six feet" (Doughty, 1974:3). With the storm Stimler's boarders increased to 40, and there was while Mrs. Butler located the Mizpah, which proved to be the richest of all. In addition, they located the Valley View, Silver Top, Buckboard, and Red Plume. Interestingly, Butler found location monuments on the property surrounding the ledges, but they had partially fallen down, looked weathered, and were apparently old; he found no location notices.

Attempts to develop the property began the second week in November when Wilse Brougher and Tasker Oddie accompanied the Butlers to the newly staked claims. Supplies consisted of mining materials, food, blankets, a water barrel, and a rickety wagon drawn by two

Leasers in front of Lottie Nay's boardinghouse. Lottie Nay is standing in the center of the doorway of the building; Mrs. Grieves is on her right. Central Nevada Historical Society – Frances Humphrey Collection

a scramble to get seats at the table nearest the fire. The storm, which lasted two weeks, brought most work in the camp to a halt. At that time there were only two 12x14-foot frame buildings and about eight small tents in Tonopah. Some of the men were without overshoes or felt boots and made use of "the company's" ore sacks for boots to walk through the snow. Because bed was the only warm place in camp, most remained there in the morning until they were sure breakfast was ready; then they crowded tightly around the stove. Stimler notes that because the floor was dirt,

Our feet would almost freeze while standing and walking on the ice and mud, while waiting on the table. By bedtime Miss Grieves' shoes and mine would be so wet that they would freeze stiff during the night. We would have to thaw them out before we could put them on, next morning. The frost would come through the canvas of the tent and drop on us, and so we had to keep our heads covered during sleep. One night I woke and found my pillow covered with light snow. Our alarm clock would freeze and stop, so I put it under my pillow to keep it warm. There were so few of us and we had so much to do that we had to get up about 4:30 in the morning. We would not get to bed until 11:00 or after at night.

Even when I got to bed I could hardly sleep; every fierce gust of wind would almost blow the tent over. Nearly every morning some tent would blow down, its occupant buried in the snow. All he could do was to pull the canvas around him to protect himself till morning (Doughty, 1974:4).

Lottie Stimler was not discouraged by the difficult conditions she first experienced in the new camp. Later she married her brother's partner, John E. Nay, and spent the chief part of her life in Tonopah.

Looking up Main Street in Tonopah, May 1, 1901. Heller Butte is on the left at the foot of Butler Mountain, previously known as Sawtooth Peak. The large building near the center is the Mizpah Bar and Grill.

old horses. As a testimony to their determination and hardiness, they brought neither tent nor stove. Initial operations were very primitive. Oddie was the youngest and least experienced in mining, so he hauled water from Tonopah Springs, made roads, and tended the camp. A windlass was used to hoist the ore. Within a short time two tons of ore had been mined and hauled 50 miles to Belmont, where freighters took it another 50 miles to Austin. There it was transferred to the railroad for a trip to Salt Lake City. The returns were eagerly awaited, and all were delighted when they received a check for $500 for the two tons. The funds enabled them to employ John Nay and John Humphrey to work in the mine and a young Indian to haul water. Oddie spent his time on horseback carrying assays to and from Belmont. Because the partners had few funds, leasers were welcomed and encouraged.

Eastern Capitalists Take Over Tonopah

Because Butler and his partners lacked the money needed to develop their claims, blocks of promising ground were at first leased to miners. But it did not take long before more

Looking northeast across the town of Butler, later to be known as Tonopah. Leases can be seen stretched out along the original discovery vein. The large building is the Mizpah Bar and Grill, where the Mizpah Hotel was later constructed. Behind it stands Lottie Nay's boardinghouse, the tent she brought from Belmont in 1901, and a small frame building she had constructed. At this point, the mines are beginning to prove themselves and there is a shift from tent structures to small frame dwellings. The chimney sticking straight out of the roof in the upper center is the Mizpah Mine hoist house.

Central Nevada Historical Society Collection

conventional methods of financing mining in Tonopah permanently changed the community. In mid-1901 Oscar A. Turner, a mining promoter from Grass Valley, California, arrived in town. Turner took an option on the properties for a price of $336,000. Oddie received $32,500 for his interests, and presumably Gayheart received a like amount. The funds were drawn on John Woodside of the American Tobacco Company in Philadelphia. The Tonopah Mining Company was established and a pattern was set for future mining operations in the area. The Tonopah Mining Company was incorporated in the state of Wyoming for 1 million shares at a par value of $1 each. Very little of the stock was offered for sale in Nevada. Jim Butler had specified to the new owners that the leasers must be allowed to work their ground until December 31, 1901, at which time the ground would become the property of the Tonopah Mining Company. The miners worked hard for the next six months trying to dig out as much ore as possible before the leases expired. Oddie became manager of the new operation.

The Greatest, the Richest, and the Best Mining Camp in the World

17

Last of the Desert Frontiersmen

Portrait by E. W. Smith of the legendary desert frontiersman, Andrew Jackson "Jack" Longstreet, circa 1905. Smith, the second photographer to come to Tonopah, operated a studio in town from about 1900 through the 1920s. He took hundreds of photos on glass negatives, many of which are in the possession of the Central Nevada Historical Society.   Central Nevada Historical Society – Fuson Collection

Jack Longstreet had a justly deserved reputation as a gunfighter with whom nobody fooled. Sally Zanjani, in her biography of Longstreet titled *Jack Longstreet: Last of the Desert Frontiersmen*, tells an anecdote which illustrates how Longstreet was perceived. It seems Longstreet was being paid $500 a month during Prohibition to let a Tonopah bootlegger with Los Angeles connections keep his still near Longstreet's ranch north of Tonopah. After an initial run with illicit alcohol to Los Angeles, law officers closed in on the canyon, split the barrels, poured liquor over the still, and set fire to the facility. Though Long-street was not home at the time, he was present the next day when the Nye County District Attorney and other law enforcement officials arrived and told him that it was necessary for them to further investigate the illicit activity. Longstreet, who appeared at the door with a long-barrelled gun, told them coldly, "Go ahead, but nobody's coming out." Zanjani says, "No one took another step forward. A consensus having very suddenly developed among all present to the effect that this particular operation required no further scrutiny, the law officers beat a hasty retreat down the canyon" (Zanjani, 1988:134).

Tonopah, late 1902.

When the Goldfield boom began in 1902, prior to the construction of the railroad linking Tonopah and Goldfield, hundreds of teams pulling wagons traveled the road between the two towns on any given day. Here a pair of ore and freight wagons can be seen having just crested the hill from Tonopah. The white portions on the hills beside the road are still visible today.

The Tonopah Railroad

W hen silver was discovered and the new mining camp began to grow, the need for transportation became acute. Myrick describes the ordeal the Tonopah-bound traveler faced:

> Passengers to the area usually stepped off the SP [Southern Pacific] trains at Reno, took the connecting V&T [Virginia & Truckee] local to Mound house, then transferred to the narrow gauge cars of the C&C [Carson and Colorado] for a six-hour ride to Hawthorne. Here an overnight layover was mandatory before catching the morning train south to Sodaville or Candelaria where stagecoach connections were available for the remaining 60-odd miles to Tonopah (1962:237).

Hauling freight was, of course, even more difficult.

There was a clamor for a railroad, but gunshy investors, who had seen mining camps come and go (in a matter of months, in many cases), wanted to be sure that this camp had staying power. By 1902, several serious schemes were being proposed for financing and building a rail line to connect Tonopah with the C&C. One project, which became the foundation of the Tonopah and Goldfield Railroad, began in February 1903, when Governor John Sparks signed a bill granting a 200-foot right of way for a railroad linking Tonopah with Rhodes, Nevada. Ultimately, financing was obtained from capitalists in Philadelphia. John W. Brock, the head of the Tonopah Mining Company, was the first chairman of the Tonopah Railroad; the relationship between the two companies continued for many years. The first "locating engineer" of the new line was an easterner who had no knowledge of the desert and its cloudbursts. The new line consequently was plagued with washouts. Bidding on construction of the line was opened in January 1904. By July 25 of that year the railroad had reached Tonopah, and a gala celebration commemorated the event.

The new company was called the Tonopah Railroad, but when it was merged with a line connecting Tonopah and Goldfield in the fall of 1905 it was renamed the Tonopah and Goldfield Railroad. The railroad was extended to Bullfrog in 1907; that leg was known as the Bullfrog Goldfield Railroad. Borax Smith, whose line, the Tonopah and Tidewater Railroad, reached Beatty after Senator William Clark's Las Vegas and

Laying the last rail into Tonopah, July 3, 1904.
Central Nevada Historical Society

Tonopah Railroad, made an agreement to use the Bullfrog Goldfield tracks until 1914, when the Bullfrog Goldfield and the Las Vegas and Tonopah tracks between Goldfield and Beatty were merged.

The Las Vegas and Tonopah Railroad was abandoned in 1918, and the right-of-way was used for the highway between Las Vegas and Beatty. The Bullfrog Goldfield leg of the T&T operation was abandoned in January 1928. The T&T Railroad running between Beatty and the Santa Fe tracks at Ludlow, California, was abandoned in June 1940. The Tonopah and Goldfield Railroad lasted several more years, servicing the needs of the Tonopah Army Air Base during World War II. It was formally abandoned on October 15, 1947.

Tonopah depot, looking
west, 1910.

Tonopah railyard, 1904. Most of the freight here was bound for Goldfield.

Looking up Main Street from about Bryan Avenue, circa 1908. The courthouse stands on the hill to the left.

CHAPTER FOUR

Slow and Steady Growth

U nlike Rhyolite and other mining communities in the central Nevada area, Tonopah grew steadily and in an unfrenzied manner between 1900 and 1910. After Jim Butler sold out there was little opportunity in Tonopah to make a fortune through speculation, and promoters left the town for nearby boom camps. Tonopah was left, as one historian put it, "to develop solely in ratio to her ore production" (Sawyer, 1931:37).

Once Tonopah began to grow, the saloons, gambling houses, and red-light establishments became the focus of social life for most miners. Business deals were discussed and plans for the area's development were made in the saloons. Saloons were the leisure-time homes for probably the majority of single miners throughout the boom period and, after working hours, the only place many operators or miners could be found. As the town grew, social life became more complex. The majority of early arrivals came from neighboring towns, and each group formed a district or neighborhood of its own in Tonopah. The districts were Little Austin, New Sodaville, and Camp Belmont. The center of town was the intersection of Main Street and Brougher Avenue. Sawyer describes the town:

Looking down Main Street one saw a number of frame shacks scattered among the white canvas tents. The Butler office stood on the lots adjoining the present Nevada National Bank building; across Brougher Avenue Lathrop and Davis had a general merchandise store; next to them was the Tonopah Club. Henry Cutting had a real estate, notary and mining office and residence in a nearby cabin. The last wooden building on the side of the street was Henry Ramsey's saloon. On the right hand of Main Street Govan and Brougher's saloon occupied the site of the present Mizpah Hotel. On Brougher Avenue, just back of this saloon, stood two tents that made up Stimler's restaurant. All of these buildings were old structures which had been hauled from Belmont and Candelaria. Tonopah had no hotel or lodging house until April, when a tent hotel was set up next to the Stimler tents. Newcomers, arriving without supplies or tents, laid their blankets on the floor of the saloon, or in some one's tent. When a woman came to camp a miner moved from his tent and loaned it to her until the arrival of her own outfit. Mrs. Butler, Miss Charlotte Stimler and Sadie Grieves were the only women in camp until April, 1901 (Sawyer, 1931:18-19).

An early view of Tonopah, (circa 1920), looking west.

Elliott states (1966:52) that as the town prospered, "the collection of dugouts, tents, and miscellaneous houses, lacking the conveniences of adequate water, fuel, and supplies, gave way" to a modern town of frame and stone buildings, daily newspapers, telephone system, the Western Union Telegraph, electric lights and power, water and sewage disposal system, railroad, ice plants, modern hotels and restaurants, and effective town government. Construction of the five-story Mizpah Hotel began in summer 1907; its doors were opened on November 17, 1908. Upper stories were completed in phases and the third story was finished in August 1909. The cost was over $200,000. A five-story building that housed a bank and various offices, with the top floors serving as apartments, was constructed diagonally across from the Mizpah.

In 1901 a post office was established in Tonopah — until 1905 known officially as "Butler City." The Nye County seat was moved to Tonopah from Belmont on May 1, 1905.

The Panic of 1907

The period from 1901 to 1907 was one of prosperity, but the three years that followed were filled with frustration and depression. By the fall of 1907, interest rates in the United States were rising and credit was tightening. A financial panic swept the nation, and depositors rushed to banks demanding their funds. Little money was available for mining development and speculation. Many small mines closed, and there was also a decline in the number of camps in the Death Valley area. Stores closed for lack of customers, and construction was halted on the new Mizpah Hotel. There was widespread fear that the Tonopah mineral deposits were in fact shallow, as evidenced by the fact that the Tonopah Mining Company's veins did not go below 700 feet. Growth in production in the Tonopah mines leveled off with the Panic of 1907, but the town weathered the storm. During the fall of 1910 there was a sudden expansion of mining activity in Tonopah, which again leveled off about 1914 and remained rather steady for more than ten years, declining precipitously in 1930.

Funeral procession for victims of Belmont Shaft fire, February 27, 1911.
Central Nevada Historical Society – William J. Metscher Collection

Fire in the Belmont Shaft – 1911

Underground mining is one of the world's most dangerous occupations. It has been estimated that 7500 men lost their lives in silver and gold mines on the western frontier. An estimate based on 1880 census data calculated that the average western miner had only an even chance of surviving to his retirement, and mine safety conditions had not greatly improved in the twenty-odd years intervening between that year and the Tonopah boom.

The Belmont Mine was Tonopah's second-richest producer and one of the most modern and well-equipped mines in Nevada, with extensive new developments in the lowest reaches of the shaft between the 900- and 1166-foot levels. But, despite its modernity, a disastrous fire occurred there on February 23, 1911. Its origins have never been determined. The tragedy took seventeen lives, but it caused less than $5000 in damages and produced no more than a slight dip in Belmont stock. The majority of the victims were foreign-born, and the average age of the fifteen men whose ages were known was only 30.

The Final Belmont Fire – 1939

Twenty-eight years after the first Belmont fire, on October 31, 1939 (Halloween night), a fire once again struck Tonopah's second-largest producing mine. An hour after quitting time, smoke began pouring out of the shaft as timbers burned. An updraft caused by the many connections between the Belmont and other shafts at lower levels intensified the smoke and fire. Sølan Terrell, who had been deer hunting that evening 40 miles east of Tonopah above the Eden Creek mine high in the Kawich Mountains, could see a column of smoke rising from the Belmont shaft. Heat from the fire and smoke warped the steel headframe, destroying the shaft buildings. The cause of the fire was never determined and the shaft was never reopened. There was no loss of life.

The majority of the ore that was being mined in Tonopah at that time was being hoisted through the Belmont shaft. Efforts were made to work through the Desert Queen shaft at the west end of the company's property, but they were never very prosperous. Total production the next year fell to less than one-third of what it had been the year before and it never recovered; the mining camp produced only a few more total tons between 1941 and 1948 than it had produced in 1939 alone. In essence, the 1939 Belmont fire was the funeral pyre for Tonopah as an underground mining camp.

Smoke pouring from the Belmont Shaft, February 23, 1911. Men standing around the shaft are waiting for the cages to come up to see who is on them and gain information on the status of the fire. The fire killed seventeen men but did less than $5000 damage to the mine and produced only a slight dip in the company's stock. Central Nevada Historical Society – Hiett Collection

Interior of the Progress
Bakery, owned and
operated by the Peter
Fabbi family, Water
Street, Tonopah,
Nevada, 1912. Peter
Fabbi is at the far left.
Central Nevada Historical Society –
Fabbi Collection

Interior of W. H. "Bill" Thomas's meat market on Main Street just west of the Mizpah Hotel, 1902.
Thomas, who was elected sheriff of Nye County in 1917, held the position for more than 35 years; he
was out of office only two years, between 1919 and 1921.
Central Nevada Historical Society – Manley Collection

Offices of the *Tonopah Bonanza*, Tonopah's first newspaper, circa 1904.

Making horseshoes in Jack Cloak's blacksmith shop, located on the southeast corner of Main and Knapp streets, circa 1905.

Mine crew, Desert Queen Mine, Tonopah, circa 1908. Note the absence of hard hats and the lunch buckets the men are holding, known as "dinner pails."

CHAPTER FIVE

In and Around the Mines

Betetween 1901 and 1921, Tonopah's mines produced 8.1 million tons of ore with a gross yield of $147,600,000. At 1989 prices for gold and silver, that would amount to more than $1 billion. Writing in 1923, Francis Church Lincoln (1982:185) declared, "Tonopah is the most important producer of silicious silver-gold ore in the United States." At that time Tonopah's production of silver was exceeded only by the large copper and lead mines, which produced silver as a by-product. In 1921, Tonopah had 4 of the 25 largest producing silver mines in the United States, and, although known as a silver camp, was the second largest producer of gold in Nevada, exceeded only by Jarbidge.

Throughout Tonopah's history, mining companies came and went. Dozens were formed during the period between 1901 and 1907, and some were consolidated into much larger firms. Most of the outlying companies folded during the Panic of 1907 and their claims became "open ground" (that is, open for relocation). In later years when ore discoveries were made near old locations, new claims were filed with new names and company owners. The mines were interconnected, often sharing tunnels and shafts. Ed Slavin, who for many years was foreman at the Tonopah Mining Company, states that there are 52 miles of drifts and crosscuts under the company's property. (A drift follows the vein and a crosscut cuts across the formation.) It is estimated that there are about 300 miles of workings beneath Tonopah. Of a vast number of companies, only a few, which are discussed here, turned out to be unusually prosperous.

The Tonopah Mining Company

The Tonopah Mining Company, the largest producer in Tonopah, yielded more than $48 million between 1902 and 1948 from two and one-half million tons of ore. Its claims encompassed all of the outcropping ledges of the district originally located by Jim and Belle Butler. A 1904 map shows about 36 shafts on Tonopah Mining Company claims, lying in a northeast to southwest direction. Many of these reached a depth of 100 to 200 feet and were the work of the first leasers.

Between 1901 and 1910 the Tonopah Mining Company led all companies in Tonopah in

CASH BOY   VICTOR   TONOPAH MINING SAND GRASS SHAFT

EXT. NEW SHAFT

(Pages 30 – 34) Panoramic view, north to south, of Tonopah, showing the location of the camp's major mines, circa 1913.

Central Nevada Historical Society – Ramsey Collection

annual production, in tonnage and in value as well as in dividends paid. During that period it produced nearly 60 percent of the town's mine production, paying out nearly one-third of its gross production values in dividends. However, between 1910 and 1930 the company's production was exceeded by the Tonopah Belmont and in some years by the Tonopah Extension and the West End. Silver dropped to 45 cents in January 1930, and the mine and mill closed down February 26, 1930. On October 3, 1930, the mine reopened under the leasing system. In 1935 there were 60 sets of leasers in the mine, but by 1944 there were only 7 left. A leasing set was a work team consisting of a minimum of two men at a particular work site in the mine. Two-man teams were required by state law for safety reasons, but the law was sometimes bent. On November 30, 1948, after forty-eight continuous years, the property ceased production. Throughout its existence, the Tonopah Mining Company paid out more than one-third of its gross profits in dividends, with production of 513,000 ounces of gold and 45 million ounces of silver, amounting to more than one-half billion dollars at 1989 precious metal prices.

The Mizpah was the deepest of the Tonopah Mining Company's shafts; it reached a depth of 1500 feet. The Red Plume went to 800 feet, the Silver Top to 700 feet, the Sand Grass to 1000 feet, and the Desert Queen to 1100 feet. The best ore in these mines occurred at the higher levels. The Mizpah mines were known as being among the dustiest in Tonopah; consequently, just as

UMATILLA

NEWGO

PAH MERGER

TONOPAH EXT.

McNAMARA

the company held the distinction of being the all-time biggest producer, its mines were also known for killing the most men by causing silicosis.

The Tonopah Belmont Development Company

The Tonopah Belmont Development Company was incorporated in 1902 in order to develop eight claims east of the Mizpah and Desert Queen shafts. Founders of the company included Tasker Oddie of Tonopah and John Brock of Philadelphia, who were also key figures in the development of the Tonopah Mining Company. Possibly because of its owners' affiliation with the Tonopah Mining Company, investors considered the firm another Tonopah Mining Company as far as its future was concerned. A discovery on the 1000-foot level in 1909 led to the development of very large ore reserves. The Belmont Vein, as the discovery was known, was quite large and in places required 40 feet of square-set and triangular-set timbering.

In 1929, mining for the company was turned over to leasers. In 1930 there were 30 sets of leasers in the mine, with a production of 4580 tons and a value of $50,208 with silver at 38 cents. This amounted to less than $1700 per set of leasers. A fire in the shaft on October 31, 1939, closed the mine, although some production continued to be associated with leasing out of the Desert Queen shaft. The mine's total production was $39.5 million from just over 2 million tons, with

almost $11 million of that, 27.5 percent, paid out in dividends. The Belmont shaft reached a depth of more than 1500 feet.

The Montana-Tonopah Mining Company

The Montana-Tonopah Mining Company was incorporated in 1902. Its organizer was Charles E. Knox of Philadelphia; Senator W. A. Clark, the Montana copper king and railroad builder, supplied much of the original capital. The company was formed to develop claims directly north of the famous Mizpah claim and around the west shoulder of Mount Oddie. One notable feature in the Montana's history is its use of leasers beginning in 1915 and continuing until at least 1923. The mine closed permanently in 1925. It produced a total of 589,000 tons, valued at $9.3 million.

The Jim Butler Tonopah Mining Company

The Jim Butler Tonopah Mining Company was organized in 1903 to develop about sixteen claims lying to the immediate south of the eastern end of the Tonopah Mining Company's property. The firm was a consolidation of small companies that had sunk four shafts on claims located close to the sidelines of the Tonopah Mining Company. The company was controlled

in Philadelphia and managed by the Tonopah Belmont. From 1910 to 1930 it was worked to a small extent by leasers. From 1903 until its closure in 1940, the mine produced 271,000 tons, with a gross value of $6.4 million.

The West End Consolidated Mining Company

Two men named Carr and Leidy were leasing on the Mizpah in 1901 when they located the West End claim. Ben F. Edwards, a borax producer from Candelaria (on property on lease from F. M. [Borax] Smith), purchased Leidy's interest in the West End and other properties. A company to promote the West End property was formed by W. J. "Billy" Douglas, Chris Zabriskie (an executive with the Pacific Coast Borax Company), and Ben Edwards, with Borax Smith advancing most of the money. Between 1906 and 1945 the company produced 774,000 tons of ore valued at about $14.5 million.

The Tonopah Extension Mining Company

The Tonopah Extension Mining Company, the third largest producer in the camp's history, was organized in 1901 to develop three claims situated southwest of the Tonopah Mining Company's easterly claims — the Sand Grass, the Red Plume, and the Buckboard, adjoining the

HALIFAX
ER TOP
BUCKEYE BELMONT
RESCUE EULER
JIM BUTLER

TONOPAH, NEV

McNamara claim. The claims originally had been located by Tom Lockhart, who had also tried unsuccessfully to locate over part of Jim Butler's claims. John McCain, a wealthy Pennsylvanian, purchased Lockhart's claims for $28,000, sunk a shaft, and encountered ore at 183 feet. McCain interested Charles E. Schwab, an industrial magnate, in purchasing a block of stock, and by 1905 Schwab was listed as owner. In 1909 a 30-stamp mill was constructed; it was eventually expanded to 50 stamps. Three of its shafts were among the deepest in Tonopah — the McCain at about 1640 feet, the Victor shaft at more than 2000 feet, and the deepest shaft in the camp, the Cashboy, at over 2375 feet — bottoming out at an astonishing 3600 feet above sea level, lower than the floor of Owens Valley to the west.

Water was first encountered at the 1170-foot level, with ever-stronger flows at deeper levels, forcing great escalation in pumping costs. A heavy flow of scorchingly hot water was discovered at the deeper levels of the Victor and Cashboy shafts. By 1926, the company was pumping about three million gallons of water a day — over 2000 gallons a minute — from the mine to the surface, with an average pumping head of about 1800 feet. Ore reserves still existed in the bottom of the Cashboy when work in the mine ceased due to the heavy flow of hot water. Operations were suspended in 1931. The Tonopah Extension Mining Company produced a total of 1.5 million tons of ore valued at $22.2 million.

A stope deep in the Mizpah Mine, 1900s, shows square-set timbers designed to keep the rock from caving in on the miners. Men with the axes are probably timbermen. Miner at far right is operating a drilling machine. The man beside the mine car may be a mucker. Note the size of the timbers at the lower right. Also note that none of the miners are wearing hard hats or using carbide lamps, working only by the light of candles.

Specialists in the Tonopah Mines

All the big dumps that can be seen around Tonopah mines were produced by hand labor. The only machines used were pneumatic drills and power hoists. There were no mucking machines or motorized trams except for electric trams used underground and on the surface by the Tonopah Extension from about 1919 to approximately 1930. Also, mules were used for tramming in the Belmont.

Generally, the low man in the mine was the mucker. He typically was given eight hours to clean out the "muck" — the newly blasted rock. In eight hours a person could load 12 to 14 one-ton cars, although men have been known to muck as many as 30 cars in a day. The mucker had to tram, or push, his car out of the face of the drift to a switch where he could get an empty car. If the drift was very wide, sometimes two men mucked at the same time.

The tasks performed by the track man were often combined with that of pipe man if the mine was small. The track man laid new track for the mine cars in the drifts and also kept old track in good working condition. The pipe man's duties involved installing and maintaining the pipes that brought water and compressed air, used to power pneumatic drilling machines, to the places miners were working. In some cases, they also involved installing large ventilation pipes that brought fresh air down to the work areas.

The timberman, another specialist, placed timbers in the tunnels and stopes. This task involved considerable skill to ensure that small rocks would not fall and large pieces of the earth would not move. Chute builders specialized in building the chutes that held the muck from stopes above prior to its being loaded into mine cars.

The machine man's duties were running the pneumatic jackhammers and stopers. It was a noisy job, and the machine man often was exposed to more dust than any of the other workers. The machine man had to be skilled in knowing how to drill the "toe or hammer cut" and he also had to decide how to drill the rock so that its unique properties or configurations could be used to blast it as effectively as possible.

There were other specialists. Some were involved with tramming the cars from the switches out to the shaft. Others loaded the full car onto the cage, or skip, once it had been delivered to the shaft by the trammer. This person was highly skilled in moving the cars rapidly on and off the

Miners drilling by hand, Tonopah, 1910.
Central Nevada Historical Society – Hiett Collection

cage and in signaling the hoist man on top according to standard code used in the mines. Old-timers say it was like poetry to see a cager one-handedly receive a full car from the trammer, spin it around, pull an empty off the cage, spin *it* around, deliver it to the trammer, put the full car on the cage, and hit the bell to the hoistman, all in one continuous motion. Any time the hoist was not operating the company was not making money, so the cager had to be as efficient as a human can be, it is said.

On top, the hoist man operated the hoist. There were trammers who took the cars once they came off the cage, trammed them to the ore bins, dumped them and put empty cars on the cage to return to the mine. When a skip was used and the ore was automatically dumped into a holding bin at the top of the shaft, it was necessary to fill cars and tram them to the large ore bins on the surface or to the dump.

Some miners worked as pipe men or track men for an entire career in the mines, preferring no other specialty. The ideal worker, however, was the all-around miner. Generally, when a man was hired in the Tonopah mines it was assumed that he could do all jobs — driller, timberman, mucker, trammer, etc.

The manager, who was in charge of the entire operation, seldom went into the mine; his duties were almost exclusively confined to the office. He took care of paperwork and frequently worked with stockholders. If the mine owned a mill, he managed it also. He knew whether the mine was paying and the various details of its economic operation. Usually he was a very competent individual; sometimes he was also the mine owner.

The superintendent was directly under the manager. A good superintendent knew most of the miners on a first-name basis and he had a good grasp of the physical operation of the mine. He knew its veins, he knew the ore, he knew where the workings were. He went into the mine whenever it was necessary for him to be there. He might check on what the ore looked like, the size of the veins, the condition of equipment, and so forth.

The foreman, who reported to the superintendent, was in charge of daily operations, and the shift bosses reported to him. One of the foreman's jobs was to supervise the quality of the ore. Generally he grab-sampled the face after every round blasted and had assays run. Before the next day's shift he knew what the round ran and in which direction the drillers should go. When the mines only ran one shift it was always arranged that the blasting would take place at the end of the shift so the smoke, gases, and dust from the blasting would not affect nearby miners.

Belmont Mine, 1400-ft. level, Tonopah, 1913.
Central Nevada Historical Society – Roy Collection

Usually each level in the mine would have a shift boss; if the level involved extensive workings, there might be more than one boss per level, particularly if there were many more than 50 men on a level. If a mine worked more than one shift, there might also be a night foreman, who usually reported to the day foreman. Often, when the mines worked more than one shift the workers would work days for two weeks, then nights for two weeks or swing and graveyard shifts for two weeks each. Shift bosses usually changed shifts following their men. The foreman did not.

Traditionally, in Tonopah, a worker was asked to complete only a required amount of production; for instance, he would muck or tram a certain number of cars. After his quota had been filled, a miner could stop working, but in most mines he could not go home. If he was allowed to go, he frequently would not be able to ride the cage to the top because they were hoisting muck; he would have to climb out. Since this required great effort, many miners would sleep in the mine until the shift was over. This unhealthy practice often led to the miner's breathing unnecessary dust and catching cold in the drafts of the mine.

Front side of a stock certificate of the 1,250,000 shares of the Tonopah Divide Mining Company at $1.00 each. Billy Douglas, a prominent mine promoter in Tonopah, promoted the mine and was the company's president. A man named Erickson was the mine's boss.

Silicosis

Dangerous as the mines were in other respects, dust was the most relentless killer of the men who worked underground in Tonopah. Tonopah rock has an unusually high silica content. When breathed into the lungs, the silica sets up an irritation that destroys lung tissue. Under a microscope one can see the ragged edges of this glassy material. The more dust a Tonopah miner breathed, the more lung tissue was destroyed. Sometimes after only a few months' work in the mines, a man became almost completely incapacitated. When a miner reached the age of forty in Tonopah — if he lived that long —it was generally considered that his life was almost over. Park benches were placed along the town's main streets so that "dusted" miners could stop and rest and catch what breath they had remaining. The two miner's hospitals that were located in Tonopah were places in which people died from silicosis. The devastating effects of the dust were well known to local residents, and many old-timers report having seen silicosis victims literally cough their lungs out in a dramatic expiration of life. The poor victims were said to have expelled masses of pink, foam-like matter from their lungs in a last dying cough.

So dangerous was the Tonopah mine dust that the many dumps in town and the dust that came home on miners' clothing endangered the entire population. Women are known to have been victims of silicosis. Norman (Curly) Coombs, a Tonopah native, describes how widespread the knowledge of Tonopah dust was. In his travels, Coombs worked in the mines in the Coeur d'Alenes of northern Idaho. In that area, miners were checked for silicosis every three years or, if they seemed at risk for the disease, every year. Coombs was checked yearly and once asked the doctor why he had to be examined every year, "when I know a lot of these guys who

Belmont mine and mill, June 1912. The mill started operation July 25, 1912.

look like they're dying take one every three years?"

"Well," [the doctor said] "it's your birthplace ... even the chickens and dogs got the God-damned con [consumption] there" (Coombs, 1990).

One physician in town, as well as numerous citizens, noted that the larger and more robust Slavic miners tended to be more susceptible to the ravages of silicosis than the smaller, more lightly built Cornish miners. The physician's explanation was that the Cornish miners' lungs were "more moist" in comparison to the Slavic miners' and that the moisture enabled them to resist the dust more — to cough it up and expel it rather than allow it to become lodged in the lungs.

Drillers, muckers, and chute-pullers were exposed to the most dust. One drift in the Mizpah had dust several inches thick on its floor. If a miner walked into the prevailing air currents, exposure to the dust was minimal, but if he went with the draft to his back, he would not be able to see his own carbide light after a short distance, so intense was the dust. The mines at the upper end of town were known to be the most dusty. The mines of the Tonopah Mining Company were the worst for exposure.

Although the dust in the mines was a horrible killer, the companies seldom made extra efforts to control it. Interestingly, the Tonopah newspapers almost never carried stories about the disease and its effects on local miners. Occasional stories in the newspapers of other Nevada communities about the disease in Tonopah were called "knockers" by the editor of one Tonopah newspaper.

Bullion ready for shipment at the Tonopah Belmont Development Company mill, March 5, 1915. Slag from the pouring of the bullion was dumped in the vessel at right, then wheeled out and dumped in slag piles.

The Leasing System

After being used briefly in the first months following Jim Butler's discovery, the leasing system in Tonopah operated from 1930 until the last mines closed after World War II. A leaser in the mining industry was comparable to a sharecropper in agriculture. (Although the technical term for a person who leases is "lessee," in mining communities throughout the western United States a person who leases and works a mining property is known as a leaser.) In the later period, the first rule in obtaining a lease was that no man could work alone underground. This rule was often ignored, but officially a person had to be able to show he would have at least one other individual with him in the area he intended to mine. The procedure for obtaining a lease was much the same for all Tonopah mines — the mines used the same lawyers in town. Some differences probably existed between Tonopah and other mining camps.

The first step in obtaining a lease in Tonopah was securing a prospector's permit, which was obtained from the mine management, cost $10, and was good for 10 days. The money went to cover the prospector's industrial insurance for the 10 days as well as the costs of hoisting him and his prospecting equipment up and down the shaft.

Each major company had a plate glass model of its mine and its veins as they were best understood. A miner was free to examine this model in order to focus his prospecting in the

The Comstock vs. Tonopah and Goldfield

There were three great precious mineral mining camps in Nevada: the Comstock, Tonopah, and Goldfield. The Comstock was in decline after the early 1880s; Tonopah and Goldfield, in contrast, were early twentieth-century boomtowns and, along with the daughter camps they spawned in southcentral Nevada, represent the last flowering of the Old West in America: the mining frontier with its individualism and the ubiquitous dream of rags to riches.

Over a sixty-one-year period, from 1859 to 1919, the Comstock produced $348 million from 11.2 million tons of ore. Tonopah, in contrast, produced $150 million from 1901 until 1948 from 8.4 million tons, and Goldfield just under $90 million from 7.74 million tons of ore between 1903 and 1947. Thus Comstock ore averaged almost twice the value of Tonopah's ore and nearly three times Goldfield's for the life of the camp. Thus, in terms of value of production, the Comstock can claim title as Nevada's greatest mining camp.

Yet Tonopah is a close second in tons produced per year. Miners estimate that it takes 12 cubic feet of unbroken rock to equal a ton. Thus, in forty-eight years of operation, Tonopah's production averaged 2,100,000 cubic feet, or 77,777 cubic yards, of ore per year. One may visualize this amount of material by comparing it to the approximately 3.25 million cubic yards of cement in Hoover Dam. Thus, from 1901 to 1948, the miners of Tonopah drilled, mucked, trammed, and hoisted to the surface enough ore from beneath the hills at Sawtooth Pass in the San Antonio Mountains to exceed by a considerable degree the volume of cement in Hoover Dam.

Production of Nevada's Three Great Camps

The Comstock, Tonopah, and Goldfield

Number of Years of Operation	Millions of Tons Produced	Value of Production	Average Tons per Year	Average Value per Year	Average Value per Ton
THE COMSTOCK 1859 -1919					
61	11.2	$348,000,000	184,000	$5.7 million	$31.17
TONOPAH 1901-1948					
48	8.4	$150,000,000	175,000	$3.1 million	$17.88
GOLDFIELD 1903 -1947					
45	7.7	$90,000,000	172,000	$2.0 million	$11.60

Figures for the Comstock are from Smith, 1980; figures for Tonopah from Carpenter, 1953; figures for Goldfield from Shamberger, 1982.

Electric-powered hoist and operator at the Halifax Mine. The Halifax shaft was 1700 feet deep.

most profitable areas. Additionally, the miner could obtain the advice of others who knew the system of ore veins. After a prospective leaser had obtained his permit, he was free to go down into the mine and look anywhere within the 52-mile system — anywhere, that is, that was not leased by someone else. The prospector was free to take samples and have them run by the company assayer at a cost of 75 cents for silver and gold. If the prospector found an available spot in the mine that he liked, he then put up $100 for supplies, including powder, caps, fuse, and compressed air. The leaser also had to pay to have his ore hoisted out of the mine. The standard lease was for 100 feet vertically and 100 feet horizontally for only the width of the vein, along with enough space to work the vein. A 100-foot cube of ground could conceivably have three or more separate leases on separate ore veins within it. As a practice, at least 50 feet usually separated sets of leasers in a work area. One could follow the vein on which he had a lease, but at the next level in the mine someone else might have a lease on it. Because the local mills were closed, ore was shipped by the Tonopah and Goldfield Railroad to the American Smelting and Refining Company in Garfield, Utah.

Under the leasing system, a leaser kept a record of the value of his ore; there could be nothing worse than to expend time and money on rock that was of no value. Most leasers shipped their ore to the smelter through a sampler. The sampler would crush the entire shipment and take a sample of it, which typically consisted of 20 pounds, 10 of which went to the smelter and 10 to the company and the miner. The smelter paid the miner on the basis of the sample. A miner could also let the smelter sample the ore, but most did not trust the smelter to do this.

The value of the miner's shipment was calculated according to the quantities of gold and silver at market prices when received by the smelter. Sometimes the market values would change while the shipment was in transit. Ed Slavin remembers shipping ore when the price was 35 cents an ounce for silver, but between the day he shipped it and the day the smelter received the ore, the price had dropped to 25-7/8 cents an ounce. Such a shift in pennies an ounce could make a huge difference in the leaser's profits.

The smelting company always took its share before paying the miner. Miners shipping Tonopah ore got a good break on their smelting charges because the smelters often needed silica, which was high in the Tonopah ores. After the refining company, the sampler took its cut, followed by the railroad. The company took its cut in the form of a royalty on the ore as well as expenses, charges for powder, fuse, and caps, and 25 cents per ton for hoisting. (In 1938, the price in Tonopah was $8 and a 10 percent handling charge for a 50-pound box of powder, $1.50 for 100 feet of fuse, and $2.25 for 100 blasting caps.) The miner had to pay his industrial

Interior of the Tonopah Belmont Development Company mill, 1912. Some of the 60 stamps operating in the mill can be seen in the background, and the giant tube mills are in the foreground. Ore from the mine first went through crushers, then on to the stamp mill and when ground sufficiently fine, on to the large tube mills.   Central Nevada Historical Society – William and Philip Metscher Collection

The Greatest, the Richest, and the Best Mining Camp in the World   43

Huge diesel engines used by the Tonopah Extension Mining Company, 1925. It is said that when the engines ran low on oil, the equivalent of what would be a quart low in oil for an autombile, a 55-gallon drum of oil was added here. The engines supposedly once powered a ship.

insurance and a "rental" on the jackhammers, stopers and steel that the company furnished. What was left went to the leaser, but if he had hired help he then had to pay his workers. During the 1930s Ed Slavin paid $6 per day to a mucker and trammer and $6.50 for a machine man (that is, a driller). Few leasers became rich, but most made a living.

Royalties to the mining company were paid on a sliding scale depending on the value of the ore. Low grade — about $15 per ton ore — carried a royalty of about 5 percent. The top royalty on high-grade was 30 percent, and anything that ran over $90 a ton was considered high-grade.

The company maintained the right to cancel a lease if a miner did not operate his ground in a "workmanlike manner." Technically, it would be hard to define a "workmanlike manner," but Slavin found (and this was the case in most mines) that no leases were cancelled, even when the leaser found high-grade. Such a cancellation, once high-grade had been found, would have spelled the end of the company, since no intelligent leaser would work with a company that defaulted on its agreements. Thus, the company had a major stake in making sure its word was good.

The leaser received what was left (if anything) after all charges and royalties were paid. Ed Slavin states:

Well, if you were lucky, it [the ore] comes out pretty fast after you get organized. You can take a four-inch streak and you can get it stripped underneath, drift on underneath and then you start stoping on it, and it can go pretty fast. Of course, it's cheaper to mine going up than going down. We always tried to get going up, but then you get into some damn places ... [He recalls one place where] ... the vein was sliding down to the north and hit something hard and pushed one piece out on the hanging wall, and then went by it and went on down. I found that piece on the hanging wall; it was 59 tons, I think, in 28 days. I'm telling you. We put timber in there, 18-inch [diameter] timbers every three feet, and they'd be crushed in the morning ... the whole damn country was moving ... Did you ever see a timber crushed? ... the rock will break, or the timber will snap and you could hear it moving, grinding, growling ... and in places that cave it will push that timber so tight an axe will bounce off of it, it makes it so hard (Slavin, 1987).

Slavin worked veins that were only 4 inches wide but ran $60 a ton. He says:

Once you start drifting on it, opening it up, but if it's good ore ... you sure don't make much ore the first month. By the end of the first month, you're in there a hundred feet and you've got that ore stripped. The next two weeks, you take out and you'll be up there another three feet, you're up nine feet and then you lag it off, and then you throw the waste down, it stays on top of that lag, and then you've got your drift underneath. But you put little chutes in, throw your ore in the chute, take it out, and the waste piles up. Every so often you have to draw it out. But you can move a lot of rock (Slavin, 1987).

Workers of extremely small veins and narrow places were known as chloriders. Ed Slavin recalls: "They would strip the vein and take the waste rock out first; they'd cob the ore off the walls with hand steel and chisel and take it out clean, so they were not paying for a lot of waste. A good chlorider is hard to find. I don't think you could find one today "(Slavin, 1987).

The Mizpah Mine, Tonopah, circa 1937.

Central Nevada Historical Society – Caine Collection

The Greatest, the Richest, and the Best Mining Camp in the World

Tonopah High School girls' basketball team, 1916. Pictured left to right are Fannie Holmes, Helen Mitchell, Ella Perry, Mae Kimball, Lavina Shields, and (front) Gertrude Rippingham.

Women in Tonopah

We tend to think of the boomtown of the Old West as wide-open socially, a place where rigid class lines are absent. This was true of Tonopah in the early days of Sadie Grieves, Belle Butler, and Lottie Stimler Nay, but in a relatively short time the situation changed. Soon, residents created class lines that were not easily crossed. This class system was very much evident among the women of Tonopah. With the exception of prostitutes and dance-hall girls, a woman's social class was usually determined by her husband's occupation and wealth, although it helped for a woman to have been born into a high social class. To a large extent a woman's choice of a husband was determined on the basis of her own social class, her wealth, whether she could put on the "airs" of a higher class, and, to some degree, her personality and initiative.

Prostitutes and dance-hall girls were at the bottom of the female social hierarchy in Tonopah. Saloons and brothels were considered debaucherous and low-status establishments. A man might enter without damage to his social standing, but women could not. Children were not permitted entry into such establishments, with the exception of newsboys selling papers in the saloons.

The next level in the social hierarchy was occupied by what might be called the "poor but decent" woman. Such women usually had little education, few job skills, and lacked the manners and refinement of women of the higher social classes. In Tonopah these women often were foreign-born. Married women who had skills and who worked outside the home often were socially stigmatized, and men whose wives worked were sometimes referred to as pimps, since a pimp lives off the earnings of his prostitute. Poor women's lives were filled with the hard work of keeping a home and caring for their family.

Poor women who became widowed (or, rarely, divorced) encountered additional hardships. In an era when no welfare was available, survival itself could be a remarkable achievement. In her book *Life of an Ordinary Woman*,

Annie Ellis describes living in central Nevada's mining camps after the turn of the century. As was true of many women of that day, she did not marry the man of her dreams but one who was available. She mourned the death of her first husband in a mining accident mainly because she had learned how "to manage him" (Ronald, 1977:96). After moving to Goldfield, her second husband lost his job, she had a miscarriage, and her daughter died of diphtheria. She was forced to look for work when her husband left her with six hungry, cold children. She resorted to stealing — not firewood or food, but a white stone from the step of a local school to make a tombstone for her daughter's grave. The stone remained at the grave site until about 1975, when it was replaced by a traditional tombstone. Women like Annie Ellis could take little joy in the present. They dealt with life's hardships by dreaming of their children's futures rather than their own. Their children were their only means to accomplishment and immortality.

In contrast, the "ladies" of the town were at the top of the social ladder. Many of them were oblivious to the lives and conditions of the lower-status residents. As members of the well-to-do, freed from the harsh realities of most women who were scrubbing, cleaning, cooking, and scraping together enough to feed and clothe hungry children and sometimes bury them, these women often romanticized their environment. They formed sewing circles and women's clubs, built and decorated lavish homes, and worked to advance worthwhile causes. "Worthwhile," of course, was defined in terms of the values of their social class and did not include a more equitable distribution of the wealth that the mines were producing. They promoted the advancement of "culture" in the community, defining culture as it would be by the wealthy in cities such as San Francisco.

Mrs. Hugh (Margery) Brown moved to Tonopah in 1904 at the age of 19, the bride of a prosperous young lawyer. The couple remained residents of Tonopah for 20 years. In her book *Lady in Boomtown*, she tells how she sent her laundry to Reno, then a three-week round trip. Remaining ever true to the principle that people see what they want to see, she described Tonopah as a "community of city people who lived in rough-board houses and walked unpaved streets, but who dressed and acted as they would in San Francisco or New York" (Ronald, 1977:94). She states:

> We all dressed with the same care we would have used in any established community. My

"calling dress" was a lovely shade of purple velvet, trimmed with an exquisite "fancy" of marabou. My traveling suit, a thin broadcloth we called "lady's cloth," was made with a short Eton jacket lined with white silk. I wore it with the red poppy hat that caught Hugh's eye at Hawthorne. The skirt was daringly short — four inches off the ground (Brown, 1968:40).

"And what of the glamorous ladies from the other end of town?" Mrs. Brown asks rhetorically in her book, speaking of the prostitutes and saloon girls, about whom she knew little. She describes how, from her husband's office in the Golden Block, diagonally across from the Butler Saloon, the short, swinging doors of the saloon permitted an occasional view in summer of a woman standing, with a foot on the bar rail, next to a man. She was intrigued by the "glamour," but knew nothing of that life.

Mining Widows

The death of many miners due to silicosis left Tonopah with a disproportionately large number of widows, many with children to support. Widowed women were forced to find ways to survive. Many no doubt left the Tonopah area, never to return. Those who remained found a supportive and helpful community where people cared about each other. Although most residents did not have much (especially after the middle 1920s), people shared what they had. The town always had a large number of bachelor miners and many widowed women provided meals and sometimes lodging for these men by running boardinghouses.

One such widow was Miruna Banovich, the mother of Catherine "Kayo" Banovich Lydon. Mrs. Banovich's husband, Mike, had migrated to the United States from Yugoslavia prior to 1900. He settled in Tonopah, sent for Miruna, the sweetheart who had waited nine years for him in the old country, and they were married in Tonopah. Banovich worked in the mines in Tonopah and died of silicosis in 1921.

Left with five children, the youngest still a babe in arms, Mrs. Banovich worked out a method of survival in Tonopah that was based on hard work and ingenuity. She ran a boardinghouse for approximately a dozen miners, and she produced a large percentage of the food she served. She kept goats in her yard, which was situated about where Main Street in Tonopah now joins Highway 6. As many as 50 goats, tended by Mrs. Banovich and the older children, grazed on the hills around the town. From the goats she obtained milk and made cheese and skorup, a type of sour cream. The

Miruna Banovich, a resourceful mining widow, pictured in the garden of her home at Magnolia and Main streets, Tonopah, during the 1940s. Mrs. Banovich, born in Yugoslavia in 1877, is holding flowers she grew in her garden.

goats — along with sheep she would buy — were also butchered, smoked, and made into kastradina, a Serbian smoked meat. In addition, she maintained a huge garden, probably the largest in town aside from that of the Lambertucci brothers west of Tonopah. She raised a variety of vegetables, including large amounts of cabbage that were used to make kraut. She had her own smokehouse for smoking meats and she kept a cow for milk and butter. Restaurant owners in town knew that Mrs. Banovich was a widow with five children to support, and they provided her with left-over fresh vegetables such as lettuce leaves and carrot tops, which the children collected daily in a wagon at the restaurants' back doors. These greens were fed to the goats and cow.

Additionally, she would obtain a ton or more of grapes each year from a wholesaler and make wine, which she sold in her boardinghouse and by the bottle. Some of the pulp from the grapes was allowed to ferment and was distilled in the still she concealed beneath the floor of the smokehouse. This produced grappa, which she also sold locally. Though it was the 1920s and Prohibition was in effect, Mrs. Banovich was always alerted ahead of time about visits from the Prohibition officers, or "Pro-his" as they were known. She was never caught.

Childhood Activities in Tonopah

Tonopah has always been a good place for a child to grow up. From its earliest days, residents placed the highest value upon the children of the town. Children were seen as the future, and the community was intent upon assuring that each child got as good a start as possible in life. The need for a school in Tonopah was recognized in early 1901. Citizens were assessed for sufficient funds to construct a temporary building, and a teacher was hired at $90 per month.

School was the main activity in a child's life. Because with a few exceptions children were not allowed to work in the mines until they were grown, most children stayed in school and graduated, although graduates were often older than today's students — sometimes 20 or 21 years of age. Ed Slavin, who moved to Tonopah in 1908 and matriculated through the local school system, says, with a twinkle in his eye, "I don't think it was because they were stupid, not as smart as today's kids; sometimes they were older because they would lay out for illness or the family would move; something like that" (Slavin, 1989).

Education from 1910 to 1940 was basic, with an emphasis on reading, writing, arithmetic, spelling, history, and geography. Because many ethnic groups lived in Tonopah, a potpourri of languages was spoken, which created quite a challenge for the teachers. Corporal punishment was allowed, and often when a child was whipped at school he could expect another whipping when he got home. School sports were popular, especially basketball. Older residents can still recall the 1925 state championship basketball team that was sent first class to Chicago to participate in a tournament.

Most of Tonopah's residents were poor. A child who wanted spending money for a movie or sweets had to earn it. Innovative and ambitious children found numerous ways to earn small sums during the 1920s and 1930s. They delivered laundry, groceries for the markets, and telegrams for Western Union. Some held jobs delivering and selling newspapers and working in restaurants and stores; others helped with housework in private homes. During Prohibition, the many boot-

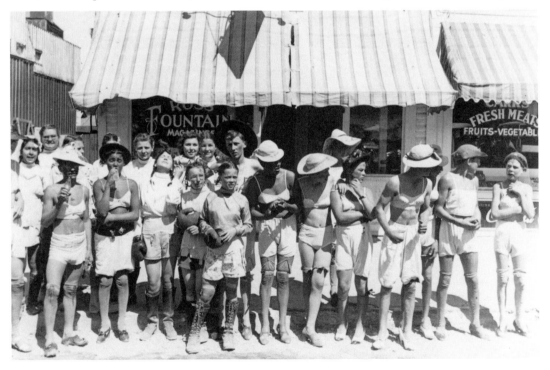

Initiation of the high school freshman class, Tonopah, September 1940. Jeanne Cirac Potts is in the back row just below the A in "fountain" in the window of Ross's Fountain on Main Street, Tonopah, Nevada. Joe Friel, wearing a large black hat, is to her immediate right. Bob Wardle is ninth from the left in the front row, wearing a wrist watch with a black band. Carr's Meat Market is at the right, and Tonopah's Butler Theatre is partially visible on the left.

Tonopah's Championship Basketball Team, 1925. Back row standing, from left, Coach Snyder, John Casselli, McGuire, Jimmy Burns, Paul Richards, and manager Bernard Fuetsch. Second row, Bill Dumble, John Starr, and Ed Slavin. Sitting in front, Chester Geyer and George Brown. While playing its last game in Ely that season the team received a telegram from home saying, "Beat Ely and Chicago next." About $2500 was raised in a few hours in Tonopah to send the team to participate in a tournament in Chicago. The Tonopah lads, however, did not fare well in Chicago against older and bigger competition.

leggers in town never had enough bottles for their wine and whiskey. Enterprising children knew who the heavy drinkers were and would visit them, collect their empty bottles, and sell them back to the bootleggers.

Burros, which ran loose in town, were a major source of pleasure and amusement for youngsters. In the middle 1920s there were 40 or 50 of the animals, descendants of burros that had been abandoned by prospectors and miners. Nobody really owned them, but children in town staked claims to individual animals. The children would ride the burros, chase them, even hook them to wagons and carts. Solan Terrell says (1987), "As mean and cantankerous an animal as was ever born was a burro. If they couldn't bite you, they kicked. If they couldn't do that, they'd roll on you." The "owners" of the burros would try to feed their animals, perhaps even stealing leftover hay from the floors of the livery stables, but most of the time the burros scavenged at the dump and around town. In fact, nobody's trashcan was safe. According to Solan Terrell (1987), "A burro is like a coyote. He can live anywhere; survive anywhere. Where any other animal would starve to death in 30 days, a burro will get fat."

There were also burros running loose on the streets of Goldfield. The youths of the two towns were mortal rivals, and Tonopah kids often would ride over to Goldfield and steal the Goldfield kids' burros; the Goldfield kids would then retaliate. Such thefts inevitably brought an attempt to retrieve the stolen animals. "If you went to Goldfield," Terrell states, "you went with a group. And you didn't slow down — you just kept moving as you went" (Terrell, 1987).

Tonopah children experienced an unusual degree of freedom; many seemed to come and go as they pleased. In the summer months, in particular, some, even those who were scarcely in their teens, would be gone for days or weeks at a time. They would take a cart pulled by a horse or burro, or perhaps ride a burro or take an old automobile, and they would move from camp to camp in the Toiyabe and Toquima mountains, fishing and enjoying themselves as only children can do. The adult community kept informal tabs on the youngsters who were out in the countryside. Ed Slavin remembers being out in the Toquimas when he was young, and being cornered by a local ranch woman; he and his friend were taken to her ranch where they were made to have a good bath. Once, Slavin's father went up in the hills to fetch young Ed because school had already begun. Mr. Slavin had no trouble locating the boys; the informal network of adults knew the exact whereabouts of the youngsters.

Fourth of July celebration and parade in Tonopah, in front of the Elks Lodge, circa 1930.

Nye County Town History Project – Fallini Collection

Children could swim in the pool filled with the warm water that was pumped from the Victor shaft at the Victor Park, at the lower end of town. Kayo Banovich Lydon describes a typical summer swim:

> We'd swim and have a good time, then the boys would decide that they wanted it [the pool] and we wouldn't give it to them. We were out there in old petticoats swimming around and the next thing you knew ... they'd throw ... carbide in the water and we'd take off. Then they'd drain the pool and clean it and fill it again so they could use it (Banovich Lydon, 1987).

The old auditorium provided space for roller skating. The children would also skate on the cement walk by the schoolhouse. In the winter there were sleigh rides down the streets of Tonopah and children would coast down Bryan Avenue on bobsleds. A ride could be made by sled from the top of Brougher Avenue all the way down to Main Street, down Main to the railroad depot, and on to the Extension Mine. Some sleds were homemade; others were store-bought.

Carnivals would come to town and set up near the present convention center. There were big parades on the Fourth of July. Dances were held at the auditorium (where the convention center sits in 1989), at the Eagles Hall, and at the high school.

Kayo Banovich Lydon remembers the good times she had as a child, herding her mother's goats on the hills surrounding Tonopah. One of her favorite activities was to climb to the top of Butler Mountain, where the old radar station presently sits. She recalls how, as small children, she and a girlfriend carved their names on a great, black rock at the top of the mountain: "Kayo — 1935" and "Sissy —1936" (the years in which they would graduate from high school). During the 1950s, when the armed forces were constructing a road to the top of the mountain, she asked one of the engineers to be on the lookout for that rock. Although he thought she was kidding, he later told her that he had found it.

Although Tonopah presented children with many opportunities for fun, it was a town of hard workers. It was natural for children to merge into the workforce once they were out of school. By at least the mid-1920s, state laws prevented children under the age of 18 from working underground in the mines, but some did so in the outlying areas surreptitiously, either working with their parents or lying about their ages. When their ages were discovered by the mine owners or by the mine inspectors, the underaged were usually sent out of the mines.

Chinese residents in Tonopah, circa 1910, in front of what was probably a laundry.

Ethnic Groups in Tonopah

E vents in Nevada's history have long been closely associated with its immigrants. In 1870, 44.5 percent of the total population of the state was foreign-born. When Bobbie Lee Colvin, now a resident of Beatty, moved to Tonopah in the middle 1930s she remembers thinking, even as a small child, that the town was extremely cosmopolitan. "You would walk down Main Street and meet Blacks, Orientals, Indians, and people from many parts of Europe speaking a variety of languages," she says (Colvin, 1988).

The Chinese were among the earliest non-Indian, non-European residents of Nye County. The census of 1870 revealed 6 Chinese living in Nye County out of a population of 1082. After Jim Butler's discoveries in Sawtooth Pass, a few Chinese moved to the new boomtown. There they established a small Chinatown complete with food stores, laundries, and lottery shops. Tonopah's Chinese were forbidden to work in the mines and instead found employment as cooks, laundrymen, and domestic servants for whites. It was "quite the thing" for Tonopah's elite to employ at least one Chinese servant.

Strong undercurrents of racism existed in most frontier mining communities. In Tonopah

on September 15, 1903, shortly after 9:00 p.m., anti-Chinese feelings erupted in violence. A racist mob descended on the Chinese section of town, rousting people from their beds; the affair left one Chinese man dead.

Although not present in large numbers in Nevada prior to 1900, Yugoslavians participated in the development of some of Nevada's earliest towns. In these communities they worked primarily as merchants, saloon keepers, and occasionally as miners. After Tonopah was founded in 1900, several individuals of Slavic origin who had moved to Nevada earlier came to Tonopah and became successful businessmen.

The south Slavs who found their way into Nevada after 1900 differed from earlier Slavic immigrants. They formed sizable communities and worked primarily as miners. Slavs in the Tonopah area were mainly Serbs from the Dalmatian coast, Hercegovina, and Montenegro. In Tonopah the south Slavs did not form a geographic enclave, but lived dispersed throughout the town. They made some effort to preserve Slavic cultural traditions. However, they were denied the status and respect accorded the first Slavic residents of Tonopah. They worked as miners and were seldom mentioned in the newspapers except in a negative light — usually some bizarre incident supporting negative stereotypes. They were hard-working people, a fact that was sometimes held against them.

Traditional customs were preserved through the celebration of Serbian Orthodox holidays. A Serbian lodge, which served as a social center for Slavic people in the Tonopah area, was also used as a church, because there was not enough money to build a Serbian Orthodox Church. Approximately 200 young Serbian miners formed the nucleus of the lodge in Tonopah.

There were many Cornish miners in Tonopah. Cornishmen have been mining the tin mines in their native Cornwall for centuries, and these men were thus skilled and highly desirable workers. In addition, a number of boardinghouses were run by Cornish women. The ethnic term used for Cornishmen was "Cousin Jacks," and the pasties they added to the cuisine of the area remain a source of enjoyment for all.

Historians have found that Italians who migrated to America often exhibited a deeply rooted desire to own land and to farm or garden. This tradition is clearly evident in the life of Victor Lambertucci. He was born in San Genesio, Province of Maceratio, Italy, and immigrated to the United States in 1907, arriving just as the financial Panic of 1907 was gripping the nation. He heard about Nevada and eventually arrived in Tonopah. In 1911, he paid $350 for a small wooden house and 15 acres of desert land to the west of Tonopah. Looking over his domain, he said, "Dis — iss goodenuf for me! I pray to my God for sendin' me here" (Ninnis, n.d.).

Noting later that a mine shaft to the east of his land had encountered water that produced a tiny stream when pumped out of the mine, he talked to the general manager of the Tonopah Merger Mining Company about the possibility of running a ditch for a farm. "Did you say a farm? This is desert lands, not farming lands," the manager replied. "Mebbe thassa what you t'ink, but sometime you comma down to my place — an' you'll see — I'ma gonna have a farm" (Ninnis, n.d.).

The manager agreed to the ditch and over the years Lambertucci developed a remarkable operation. Between 1921 and 1940 over 3700 head of hogs were butchered and sold at his establishment. A hothouse 136 feet long and 34 feet wide was constructed. During the 1930s Lambertucci dismantled the old United Cattle and Packing Company's packing house in Goldfield and moved it to his property. Victor's brother Dominic joined him at the farm.

Fields and buildings, including a slaughterhouse, on Victor Lambertucci's property on the west side of Tonopah, circa 1910. Lambertucci obtained water from the Victor and Tonopah Extension mines for his crops. He grew grapes and made wine, which was sold locally.   Central Nevada Historical Society – Fuson Collection

Interior of the Big Casino, circa 1908. The Big Casino featured the highest quality orchestras and bands from Reno and San Francisco; old-timers say the entertainment was the quality of anything found in Las Vegas today. Men paid female employees of the establishment to dance. A brothel was located on the second floor. Early in the evening the band would play outside on the sidewalk; this was known locally as "the call of the wild," which signaled that the casino was open and ready for business.

CHAPTER SIX

Tonopah's Nightlife

rom its earliest days Tonopah has had an active nightlife. The bars always constituted one of the town's major social centers. Aside from the Mizpah Hotel, the Big Casino, which opened April 1905, was probably Tonopah's most famous establishment. It was located on the east side of Main Street between Oddie and Knapp avenues. The Big Casino contained a restaurant, saloon, betting hall, and a dance hall. Cribs for prostitution were located on a balcony above the dance floor. Men paid to dance with women employed by the establishment, who would get their partners to go to the bar and buy drinks. The girls worked on commission and received 40 percent of all drinks they sold and 50 percent of all dance receipts. Dance girls were not necessarily prostitutes; a man went to the balcony for sexual encounters with these women.

By 1907 the Big Casino boasted a small orchestra. Entertainers were imported from San Francisco and were of a quality to equal Las Vegas today, remembers Ed Slavin, who sold newspapers in the Big Casino as a boy. The Big Casino had direct wires to the major racetracks in the country and made book on important sporting events. It claimed to be the largest hurdy-gurdy house on the West Coast at the time. In 1907 it helped sponsor the Gans-Herman World Lightweight Championship fight. With the enactment of antigambling laws in 1910 it fell on hard times and by 1913 was in financial trouble; it went into receivership, a litigant in the federal courts.

By 1920 the dance hall had been abandoned and the largest of Tonopah's "joints," as they were always called, had been converted into the Big Casino Hotel. On August 23, 1922, the "Monte Carlo of the Desert," famed in song and glorified in literature by Jack London, Rex Beach, and Robert Service, was destroyed in a $75,000 fire, which razed the entire lower end of downtown north of Oddie Avenue.

Tonopah's red-light district was bounded by Main Street and Knapp, Oddie, and Central avenues. There were two kinds of establishments in the district — bars and dance halls, and brothels. Old-timers in town say that during its heyday there might have been as many as 300 dance-hall girls and an equal number of prostitutes in the district. One long-term Tonopah resident, when asked what the town's nightlife was like during the 1920s, replied, "Tonopah

Big Casino Hotel, circa 1917. Looking up Main Street, with the Mizpah Hotel visible in the background. The Big Casino was originally a gambling establishment, dance hall, and brothel. When gambling was outlawed, the Big Casino fell on hard times and was forced to convert to a hotel.

Central Nevada Historical Society – Nevada Historical Society Collection

Interior of Stock's liquor store, Tonopah, Nevada, 1907. The establishment featured barrels of wine and whiskey. The bar had a brass rail and a well-placed spitoon, but no stools.

Central Nevada Historical Society – Pietrykowski Collection

Interior of the Tonopah Club, located on Main Street, circa 1910. The presence of "respectable" women in the photo suggests a special occasion, as they were ordinarily not present in saloons. The cashier's cage is in the background at right.

Central Nevada Historical Society – Slavin Collection

doesn't have a nightlife now in comparison" (Skanovsky, 1987). Miners tended to be hard-living individuals; they worked hard in the mines and pursued their nighttime activities with equal vigor. Men would be seen on the streets at night in an inebriated state, barely able to walk, and yet the next day they could put in a shift at the mine with no hint of the previous night's activities.

During the 1920s, many of the establishments featured live music, often top groups from San Francisco and Reno. Free hors d'oeuvres and other food items were frequently available. A person could get a good sandwich of pork, beef, or turkey and a beer for 5 cents. Like so many business establishments in town, the bars and dance halls in the red-light district had their own tokens with which a patron could purchase goods and services. Tokens from one place were good throughout town, to be used as if they were cash. Most establishments' tokens came in 6-1/4- and 12-1/2-cent denominations, with the former substituting for a nickel and the latter for a dime. Thus, patrons received a value bonus to encourage further purchases. If a patron went into an establishment and ordered a 5-cent beer and sandwich and paid for it with a dime, he would receive a 6-1/4-cent token in change, thus receiving a 1-1/4-cent bonus. After the establishments folded and the tokens no longer had purchasing power, many townspeople threw them away. Many ended up in the dump west of Tonopah, where some were retrieved years later by souvenir and artifact hunters. Collections of these tokens are now prized by their owners.

Although women in the red-light district were plentiful, and many were beautiful and could promise a good time to anyone who could pay the price, there was always the risk of

Bobbies Bar (brothel), St. Patrick Street, Tonopah, circa 1945. Bobbie Duncan, pictured fourth from the left, later established the Buckeye Bar on U.S. Highway 6, just east of Tonopah.

venereal disease. Syphilis and gonorrhea were not uncommon among the red-light district patrons, and in that pre-antibiotic era the diseases were troublesome and dangerous.

By the late 1930s economic activity in Tonopah was a shadow of its former self. Despite this decline, there were still more than a dozen bars and brothels in the red-light district. These included the Silver State Bar, Effie's Place, the Lucky Strike, Fay's Place, the Green Lantern, the Bungalow, Taxscine's Place, the Northern, the Newport Dance Hall, the Cottage Bar, Inez Parker's Place, and Nigger Dee's.

Tonopah's red-light district was still in operation in the early 1940s when the United States Army Air Force arrived. During the construction phase of the air base at the beginning of World War II, all the joints in Tonopah closed at midnight so the workers would arrive at work free of hangovers and generally prepared for the next day's hard work. But when a fire destroyed one of the houses, killing one soldier, the army complained. It was instrumental in having the red-light area shut down. Several of the houses reopened in the 1950s but not on the scale of previous years.

Prostitutes in Tonopah have had a long-standing reputation of being generous to others in the community in times of need and tragedy. They have been known to bankroll many a needy cause and to serve as surrogate mothers to untold numbers of young men. Girls from the red-light district sometimes even helped Tonopah's youth with college expenses.

Interior of the Ace Club, Tonopah, late 1930s or very early 1940s, certainly pre-World War II. Louis Cirac, wearing the sweater, father of Jeanne Cirac Potts, is dealing roulette. The man to his right, raking in chips, was named Palooka.

Nye County Town History Project – Potts Collection

Freck Lydon, husband of Catherine Banovich Lydon, and Jack Dempsey, former World's Heavy-weight Boxing Champion, at Tonopah's 50th-year celebration in 1950. Dempsey was the Grand Marshal in the Tonopah 50th Anniversary Parade. Dempsey knew Lydon from his early days in the Tonopah-Goldfield area, when Dempsey roomed (batched) with Lydon's brother. Commenting on his fights with a local fighter named Johnny Sudenberg, Dempsey said, "I fought Sudenberg twice — once in Goldfield and again in Tonopah. They were the two toughest fights of my entire career." Both were declared draws. Recalling the first encounter held in Goldfield in 1914, Dempsey said, "I don't remember anything after the 7th or 8th round. I woke up in my room the next morning, the fight was over and they told me it was a draw" (*Goldfield News*, May 26, 1950). Freck Lydon had trained with Sudenberg and Lydon was involved in law enforcement in Tonopah for many years. If a trouble-maker saw Lydon put on his gloves when making an arrest, the troublemaker knew he was in for a tough time!

Tonopah, Nevada, looking southwest, circa 1940. Water tanks on the hill near center were placed there in 1915 following disastrous fires the previous year that burned entire sections of Tonopah because water pressure was lacking. The high school stands at the upper left center, looking up Bryan Avenue.

Ruby Keplinger Moore and Roy Moore, Rex Bar, Tonopah, October 1946.

P-39 Airacobras and hangar at the Tonopah Army Air Base, 1943. Later that year, the army replaced the P-39s at Tonopah with B-24s due to numerous crashes of the P-39s in the area.

CHAPTER SEVEN

The Tonopah Army Air Base

For Tonopah, 1940 was a watershed year. Mining production had dropped precipitously in 1930, risen gradually between 1931 and 1937, leveled off until 1939, and then, following the second Belmont fire, plunged to deep Depression levels in 1940. Underground mining production never recovered in the Tonopah area. Ironically, in the year the old mining-based economy died, a new economic force emerged for the community. In 1940, the U.S. Army began construction of the Tonopah Army Air Force Base about 7 miles east of town, marking the beginning of a dependence on the United States government and its military for the community's economic well-being.

After Germany attacked Poland in September 1939, General Headquarters Army Air Force at Langley Field, Virginia, began considering the desirability of improving the airdrome at Tonopah and obtained a large tract of land for gunnery and bombing practice. In October 1940, approximately 5000 square miles of land, roughly 4 percent of the state of Nevada, was transferred from the Department of the Interior to the War Department. Construction of a new airfield began in early 1940, sponsored by the Civil Aeronautics Administration and financed in part by the Works Progress Administration. Condemnation proceedings on private holdings within this vast area were instituted and resolved in August 1941.

The base, which was ready for occupancy by July 1942, came under the jurisdiction of the Fourth Air Force and would be used for training purposes. Facilities included runways, mess halls, and a hospital. By January 1943, 1779 enlisted men and 227 officers were stationed at the field. Training was provided on the Bell P-39 Airacobra, which was being used in the Pacific by the U.S. Army Air Force. Because of the high loss ratios of both planes and pilots in crashes at the base (possibly because of the high elevation of the base or design problems with the planes), in mid-1943 it was decided that the base would be used to train crews of B-24 Liberators.

In November 1943, the Air Force began $8 million worth of improvements on the facility. These included construction of a concrete apron 600 feet wide and 1 mile long; two main runways 150 feet wide and a total of 4000 feet in length; two taxiways totaling 10,000 feet; a reinforced concrete water storage reservoir with a capacity of 1 million gallons; approximately

Enlisted Men's Service Club, Tonopah Army Air Base, about 1943.

140 buildings of all types and sizes; a complete sewage disposal system; and an 8-inch, 14-mile line of pipe to supply water from wells at Rye Patch.

On average, 1000 men working one shift a day were employed on this part of the construction. There was a shortage of skilled labor, but unskilled recruits were soon taught the necessary skills. Prior to the installation of booster pumps at Rye Patch, it was necessary to truck water for the project from Millers, 27 miles away. In September 1943, half the personnel stationed at the base were moved temporarily to Bishop Army Air Field in California so there would be enough housing for the large numbers of construction workers. Additional quarters and barracks, a new post exchange, supply buildings, day rooms, crash stations, warehouses, operations buildings, a hangar, and a school building were also constructed. By November 1943, most of the work had been completed and the men who had been sent to Bishop returned.

In the summer of 1944, in a move that was to preview future uses of the Nevada desert, a special weapons test organization from Wright Field, Ohio, tested glide bombs and other devices at Tonopah. The training on B-24s continued until the end of the war; by October 1944, there were 66 B-24s in use in that training program, and there were 5273 enlisted men and 1264 officers stationed at the base in addition to a large number of civilians. By March 1945 the number of personnel was down to 3707 enlisted men and 437 officers. A week after the fighting ended in the Pacific, the Fourth Army Air Force placed the Tonopah base on inactive status, and in 1948 the base was deactiviated and materials sold for scrap. Runways and four hangars in various stages of disrepair are all that remain of the base.

When Tonopah residents first heard that the government planned to build an air base nearby, they could scarcely believe it. Tonopah was a quiet little town. "What a lot of malarky," they thought. The next thing they knew, the air base was under construction. On-site construction provided jobs for local residents, and the influx of large numbers of workers provided a major stimulus to the town's economy. Anyone who was willing to work could find a job, either on the site or in the bars, restaurants, and other business establishments serving the workers in town. The bars and gambling establishments, including the Tonopah Club, the Ace Bar, the Coors Bar and the Mizpah Hotel, did overflow business. There was prosperity for the town's merchants. Downtown was described as a "crazy house" and the restaurants often had standing-room-only crowds. Service personnel were housed as far away as the Goldfield Hotel. Many local families took on boarders or opened their homes to service men and their families. Housing conditions were generally poor, with people living in anything they could get, including refurbished garages and chicken coops. Conditions were particularly primitive for black families, who lived in a segregated section on the west side of Water Street and west of Corona Avenue.

With so many single men in town, unmarried local girls were at a premium. Any girl could take her pick and have as many boyfriends as she wanted. During this period, many close friendships that continue today were forged between Tonopah residents and the airmen. Local boys, however, often resented the airmen "stealing" their girls.

Closure of the air base at the end of the war produced a rapid decline in the town's economy. Mining activity remained small and sporadic, and tourism was still largely undeveloped. People who remained in Tonopah had to have survival skills. If one did not own a ranch, a profitable commercial establishment, or find some government employment, a mixture of methods had to be used to keep a roof over one's head. This mixture usually involved mining, day labor and odd jobs, and when possible, some type of entrepreneurial activity.

Aerial view of the Tonopah Army Air Base, January 1944 , looking south toward Mud Lake. Highway 6 is at the lower left corner; runway and taxiways can also be seen. Keyhole-shaped markings near the center are parking spots for P-39 Airacobras, each equipped with electric hookups for the aircraft.

Central Nevada Historical Society – Jacobsen Collection

The Hood Test, conducted at the Nevada Test Site on July 5, 1957, was a 74-kiloton device exploded from a balloon. The author viewed this test from the Reveille Mill located in Reveille Valley north of the Test Site.

CHAPTER EIGHT

The Tonopah Test Range and the Nevada Test Site

After the closure of the Tonopah Army Air Base and liquidation of most of its facilities in 1948, there was sporadic activity at the military facilities east of town. During the late 1950s, this activity consisted of construction of housing for high-speed cameras, control facilities, and large concrete pads for bomb-drop tests. Some badly needed jobs were provided for the community, but the overall impact was not great. It was during this period that radar stations were constructed atop mountains to the immediate south and to the north of Tonopah. Although the facility atop Mount Brock to the south was scarcely used by the air force, the radar site to the north necessitated stationing several airmen in Tonopah.

In Spring 1949, government officials stated that it would take a national emergency to justify atomic testing within the borders of the United States. Such an emergency arose in the summer of 1950 when the United States became involved in the Korean conflict, shortly after the detonation of an atomic bomb by the Soviet Union. The site selected for testing was part of the old Las Vegas-Tonopah Bombing and Gunnery Range located in Nye County. The first series of tests, code-named "Operation Ranger," took place between January 27 and February 6, 1951. Initially, tests were conducted in the atmosphere; by the time the limited test ban treaty was signed in Moscow on August 5, 1963, all tests had been moved underground.

Between 1951 and 1989 approximately 700 nuclear devices were detonated at the Nevada Test Site. This facility was a major factor in the economies of Nye and Clark counties from the 1950s through the 1980s—in the mid-1980s, involving either directly or indirectly about nine percent of the workforce in southern Nevada.

Tonopah area residents have vivid memories, both positive and painful, of atmospheric testing of atomic weapons. During the 1950s, most of the atomic shots were detonated in the hour before dawn. Usually, they were announced in advance on the radio. Residents of Tonopah sometimes drove to Salisbury Wash and the Warm Springs Summit on Highway 6, east of Tonopah, to watch the atomic explosions in the pre-dawn sky. They can recall vividly the flashes of beauty and the awe they felt for the incredible power from splitting the atom. I recall watching many such explosions on pre-dawn mornings when living at the Reveille Mill during the 1950s. My most vivid memory concerns a mid-day shot, however. We were working on repairs at the old mill and took a noon break in the shack to eat and escape the heat. Words cannot describe our astonishment when we exited the shack in the early afternoon: The door faced south toward the Test Site, down the Reveille Valley, and in the direction of Reveille Peak, clearly visible, a perfectly formed mushroom cloud was rising above the horizon.

During the late 1950s, additional jobs opened up for miners on the Nevada Test Site as the government shifted from atmospheric to underground testing of atomic weapons. Long tunnels were drilled under Rainier and Aqueduct mesas and nuclear devices were exploded; the radiation and other effects of the blast could be better contained underground. During the summer of 1958, there was a large expansion of the tunnel-drilling force at the Test Site. The camp at Area 12 was established. Miners who did not wish to make the long commute from Las Vegas, Amargosa Valley, Pahrump, or Beatty stayed at the camp and returned home on weekends. Many miners considered the camp at Area 12 home for more than 10 years — in some cases, closer to 20. Other workers came and went. A person could work at the Test Site for a time, build up some savings, and then go prospecting or mining. Later, when his funds were low, he could return to the Test Site. Many of the older miners and construction workers employed at the Test Site who were used to working on jobs where profit margins were small to nonexistent were critical of the waste of material and labor that they saw at the Nevada Test Site.

Some atomic devices were exploded in horizontal tunnels, others at the bottom of deep vertical shafts. The vertical shafts, 3 to 12 feet in diameter, could be several thousand feet deep. Sometimes tunnels were driven out from the bottom of the shafts, other times the shaft bottoms were enlarged. After the atomic device was blasted, there was a recovery stage, which involved drilling into the blasted area to determine effects on equipment, instruments, and the rock.

During the 1970s and 1980s activities on the Tonopah Test Range expanded and provided additional employment opportunities. Expansion became especially pronounced during the 1980s during the military buildup of the Reagan administration. A squadron of Stealth fighters was based on the range.

The Proposed Yucca Mountain Repository

Since the U.S. Congress passed the Nuclear Waste Policy Act in 1982, there has been a great deal of activity by the U.S. Department of Energy (DOE) to determine the geotechnical suitability of Yucca Mountain to be the nation's first high-level nuclear waste geologic repository. Yucca Mountain is located approximately 15 miles southeast of Beatty, Nevada, on land controlled by the federal government. The facility must be designed to safely isolate large quantities of highly toxic and dangerous nuclear waste from the human environment for 10,000 years, and the DOE will not know if the location is suitable until site characterizations are completed in approximately 1996.

If the site is judged to be geotechnically suitable, the DOE must then receive a permit to construct and operate the proposed nuclear repository from the Nuclear Regulatory Commission. Under even the most optimistic scenarios, it is not expected that a repository would be operational until the first decade of the twenty-first century.

Resentful of Nye County's effort to actively participate in the federal government's repository siting program, and desirous of receiving federal funds earmarked for Nye County, the 1987 Nevada Legislature removed from Nye County's jurisdiction a 12-mile-square block of land containing Yucca Mountain and created a new county called Bullfrog County (which

Large diameter bits are used to drill "big holes" at the Nevada Test Site. Big holes measured 36 to 144 inches in diameter with depths from a few hundred to five thousand feet. Miners from the Tonopah area often worked at the Test Site on these drilling rigs. Atomic devices were detonated at the bottom of the drill holes.

Central Nevada Historical Society – DOE Collection

would have no residents). The name related to an unsuccessful attempt to create a new county (Bullfrog) out of southern Nye County during the heyday of Rhyolite in 1906. Nye County asked the district court to rule the law creating Bullfrog County a violation of Nevada's Constitution. In February 1988 the district court affirmed Nye County's position; hence, Bullfrog County was short lived (Bradhurst, 1988).

Late in December 1987, the U.S. Congress passed legislation that superseded the 1982 Nuclear Waste Policy Act and singled out Yucca Mountain as the prime candidate site for storage of high-level nuclear waste. Supporters of the 1987 act agree that the government's choice of Yucca Mountain for the possible storage of high-level nuclear waste is sound, given the area's aridity and sparse population, the large amount of nuclear testing that has taken place at the Nevada Test Site over more than 35 years, and the necessity of restricting the area for thousands of years to come regardless of future waste storage projects.

Those opposed, on the other hand, point out that there is a big difference between the relatively small levels of waste produced through nuclear testing and the large amounts planned for storage. They argue that most of the waste is produced in the eastern United States, and that Nevada has done enough for the country with the atomic testing program and its many military installations. Moreover, they fear the unknown problems such a facility might present. The new law, however, does seem to make a Nye County site for a high-level nuclear waste facility a strong possibility, since it focuses on Yucca Mountain.

From left, Allen, Philip, and William Metscher in front of the Central Nevada Museum, Tonopah, 1990.

The Central Nevada Museum

Residents of Central Nevada have long been aware that the area has a unique and interesting history, well worth preserving for future generations. In 1978, area residents organized the Central Nevada Historical Society. One of their goals was to establish a museum in Tonopah. Bylaws were created, a dues system was placed in effect, and the organization held its first and only membership drive by sending out a mailing from a list of subscribers to the *Tonopah Times*. The organization obtained some 400 members from this mailing and membership remains at about 450 to 500. The society began a publication entitled *Central Nevada's Glorious Past* that is published twice a year and features pictures and stories about the area. The group began to collect artifacts and place them in storage.

Soon thereafter, the Fleischmann Foundation, a Nevada-based organization, awarded the Central Nevada Historical Society (through Nye County) $215,000 for the construction of a museum. This structure stands between the hospital and the power company off Main Street on the east side of Logan Field Road. The grand opening was held in July of 1981. William, Philip, and Allen Metscher have been the driving forces throughout the formation of the historical society and the development of the museum. The museum is a must-see for every Tonopah visitor.

References

Banovich Lydon, Catherine. *An Interview with Catherine Banovich Lydon.* Nye County Town History Project, Tonopah, Nevada, 1987.

Bradhurst, Stephen T. Personal communication. 1988.

Brown, Mrs. Hugh. *Lady in Boomtown: Miners and Manners on the Nevada Frontier.* American West Publishing Company, Palo Alto, California. 1968.

Carlson, Helen Swisher. *Nevada Place Names: A Geographical Dictionary.* University of Nevada Press, Reno, Nevada. 1974.

Carpenter, Jay A. "The History of Fifty Years of Mining at Tonopah, 1900-1950." In "The History of Fifty Years of Mining at Tonopah, 1900-1950." Jay A. Carpenter, Russell Richard Elliott, and Byrd Fanita Wall Sawyer. University of Nevada *Bulletin,* pp. 43-153. January 1953.

Carpenter, Jay A., Russell Richard Elliott, and Byrd Fanita Wall Sawyer. "The History of Fifty Years of Mining at Tonopah, 1900-1950." University of Nevada *Bulletin.* January 1953.

Cline, Gloria Griffen. *Peter Skene Ogden and the Hudson's Bay Company.* University of Oklahoma Press, Norman, Oklahoma. 1974.

Colvin, Bobbie Lee. Personal communication. 1988.

Coombs, Norman "Curly." A Chart of the Depth of Mines in Tonopah, Nevada (in Mr. Coombs's possession; Tonopah, Nevada). Circa 1928.

Clyde Rufus Terrell, owner of the *Tonopah Times Bonanza*, seated at the linotype machine on which he would compose stories for the paper (circa 1948). Nye County Town History Project – Terrell Collection

————. *An Interview with Norman Coombs.* Nye County Town History Project, Tonopah, Nevada. 1990.

Doughty, Nanelia S. "Lottie Nay — Tonopah Naught One: Bitter Winter Storm Lasted Two Weeks," *Las Vegas Review Journal* "Nevadan," pp. 3-5. June 2, 1974.

Elliott, Russell Richard. *Nevada's Twentieth Century Mining Boom: Tonopah, Goldfield, Ely.* University of Nevada Press, Reno, Nevada. 1966.

Ellis, Anne. *The Life of an Ordinary Woman.* Houghton Mifflin, Boston. 1929. Cited in "The Tonopah Ladies" by Ann Ronald, *Nevada Historical Society Quarterly,* Vol. 20, No. 2, pp. 93-100. Summer 1977.

Elston, Robert G. "Prehistory of the Western Area." In *Great Basin,* ed. Warren L. D'Azevedo. *Handbook of North American Indians,* Vol. 11, pp. 135-148. Smithsonian Institution, Washington, D.C. 1986.

Funk, Ann W. "Peter Ogden: Nevada's First Great Explorer," *Las Vegas Review Journal* "Nevadan," pp. 6L-7L. May 30, 1982.

Goldfield News, May 26, 1950.

Keeler, P. E. "Nye County." In *The History of Nevada,* Sam P. Davis, ed. Elms Publishing Co., Los Angeles, California. Pp. 960-972. 1913.

Interior of Safeway store in Tonopah, 1938. The store was sold and later became Coleman's.

Lincoln, Francis Church. *Mining Districts and Mineral Resources of Nevada.* Nevada Newsletter Publishing Company, Reno, Nevada. 1923. Reissued by Nevada Publications, Las Vegas, Nevada. 1982.

Myrick, David F. *Railroads of Nevada and Eastern California.* Vol. 1. Howell-North Books, Berkeley, California. 1962.

Ninnis, Lillian. "The Story of Victor Lambertucci," a series of articles appearing in the *Tonopah Times-Bonanza and Goldfield News.* On file at Central Nevada Museum, Tonopah, Nevada. File No. 36. n.d.

Ronald, Ann. "The Tonopah Ladies," *Nevada Historical Society Quarterly*, Vol. 20, No. 2, pp. 93-100. Summer 1977.

Sawyer, Byrd Fanita Wall. "The Gold and Silver Rushes of Nevada, 1900-1910." Unpublished M.A. thesis, University of California. 1931.

———. "Extracts from 'The Gold and Silver Rushes of Nevada — 1900-1910.'" In "The History of Fifty Years of Mining at Tonopah, 1900-1950." Jay A. Carpenter, Russell Richard Elliott, and Byrd Fanita Wall Sawyer. University of Nevada *Bulletin*, pp. 1-17. January 1953.

Shamberger, Hugh A. *Goldfield.* Nevada Historical Press, Carson City, Nevada. 1982.

The Reveille Mill, located in Reveille Valley east of Tonopah, as it looked in 1926.

Shepperson, Wilbur. "Immigrant Themes in Nevada Newspapers." *Nevada Historical Society Quarterly*, Vol. 12, No. 2, pp. 5-46. Summer 1969.

Skanovsky, Bruno, and Ann Skanovsky. *An Interview with Bruno and Ann Skanovsky.* Nye County Town History Project, Tonopah, Nevada. 1987.

Slavin, Edward R. *An Interview with Edward R. Slavin.* Nye County Town History Project, Tonopah, Nevada. 1987.

———. Personal communication. 1989.

Smith, Grant H. "The History of the Comstock Lode, 1850-1920." University of Nevada *Bulletin*, Vol. 37, No. 3. July 1, 1943. Ninth printing. 1980.

Terrell, Solan. *An Interview with Solan Terrell.* Nye County Town History Project, Tonopah, Nevada. 1987.

Thomas, David Hurst. "The Colonization of Monitor Valley, Nevada," *Nevada Historical Society Quarterly*, Vol. 25, No. 1, pp. 2-27. Spring 1982.

Tonopah Bonanza, April 29, 1905.

Zanjani, Sally Springmeyer. *Jack Longstreet: Last of the Desert Frontiersmen.* Athens, Ohio University Press, 1988.

Like a ghostly reminder of glorious days past, the headframe of the Far Western Mine, west of Tonopah, stands amid the desert's eternal splendor. The Malpai Mesa is visible in the background, and Lone Mountain looms in the distance, its spirits having seen the coming of Native Americans to the Great Basin; the booming, then fading, of Tonopah as a mining camp; and the survival of Tonopah as a mining, ranching, and tourist center.

About the Author

Robert D. McCracken, a descendant of three generations of hardrock miners, was born in the high country of Colorado, where he lived until he was eight. His love for Nevada and its people began in the 1950s when he and his brother helped his father operate mines at several sites in Nye County, including Reveille Valley and Silver Bow. During his college years, McCracken worked in Nye County on construction jobs. He earned his Ph.D. in cultural anthropology at the University of Colorado and has taught at Colorado Women's College, California State University at Long Beach, and UCLA. He is the author of numerous scientific reports and articles and was cited in *Time* for his work on human evolution. In 1981 he returned to Tonopah, where his father had retired. He began the Nye County Town History Project in 1987.

Books from Nye County Press

by Robert D. McCracken

A History of Amargosa Valley, Nevada (cloth)
ISBN: 1-878138-56-1

The Modern Pioneers of the Amargosa Valley (paper)
ISBN: 1-878138-58-8

A History of Beatty, Nevada (cloth)
ISBN: 1-878138-54-5

Beatty: Frontier Oasis (paper)
ISBN: 1-878138-55-3

A History of Pahrump, Nevada (cloth)
ISBN: 1-878138-51-0

Pahrump: A Valley Waiting to Become a City (paper)
ISBN: 1-878138-53-7

A History of Tonopah, Nevada (cloth)
ISBN: 1-878138-52-9

Tonopah: The Greatest, the Richest, and the Best Mining Town in the World (paper)
ISBN: 1-878138-50-2

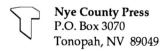
Nye County Press
P.O. Box 3070
Tonopah, NV 89049